Self-Preservation at the Center of Personality

Superego and Ego Ideal in the Regulation of Safety

Ralf-Peter Behrendt

Vernon Series in Cognitive Science and Psychology

VERNON PRESS

www.vernonpress.com

In the Americas:
Vernon Press
1000 N West Street,
Suite 1200, Wilmington,
Delaware 19801
United States

In the rest of the world:
Vernon Press
C/Sancti Espiritu 17,
Malaga, 29006
Spain

Vernon Series in Cognitive Science and Psychology

Library of Congress Control Number: 2016953832

ISBN: 978-1-62273-120-6

Cover design by Vernon Press, using elements selected by freepik .

In memory of my mother,
Amanda Behrendt,
1939-2016

Table of Contents

Introduction

Heinz Kohut (1971, 1977) developed self psychology, a branch of psychoanalytic theory, in recognition of the central role of self-esteem and self-cohesion in the functioning of the personality. Kohut did not expand on the resonances in his theory with the works of Alfred Adler, Paul Federn, Karen Horney, and Joseph Sandler; but neither did Sandler highlight the contributions of Horney and Federn, or Horney those of Adler and Federn, yet their theories are highly compatible and complement each other. The 'principle of self-preservation', advanced by self psychology as the fundamental principle underlying social behavior and personality organization, stipulates that the subject must maintain his ties to his selfobject surround if he is to preserve the integrity of his self (Stolorow, 1983; Brandchaft, 1985). In other words, the personality is organized around the need for *approval* (Flugel, 1945), specifically, and the need for safety (Sandler, 1960a), more generally. This imperative can be in *conflict* with other demands, internal ('instinctual') or external. All psychological conflicts are ultimately concerned with the preservation of the integrity (cohesion) of the self (Stolorow, 1985). Ego defenses, the focus of classical psychoanalysis, are 'ego functions' (Hartmann) that serve the preservation of the self (ego), that is, the subject's sense of connectedness to the selfobject surround (and hence his feeling of safety [Sandler]). Ego defenses (defense mechanisms) resolve conflicts between the need for safety and 'instinctual drives' (drive impulses). Drive impulses arouse anxiety (and hence are consciously intolerable) insofar as the resulting behavior would be socially inacceptable (and invite disapproval) and would thus threaten the narcissistic homeostasis and integrity of the self. The self is 'narcissistically cathected' (Hartmann, 1964; Jacobson, 1964), meaning that it is constituted, and maintained in its cohesiveness (Kohut), by others' approving attitudes toward oneself and by others' recognition and acceptance of oneself, attitudes that are induced and have to be maintained by oneself through employment of what can be called 'narcissistic behaviors' (proximally concerned with others but ultimately with oneself and one's safety). 'Self' and 'ego' are treated synonymously in this book, in keeping with Freud's earlier work and also with Federn (so that, for the most part, 'ego' here is *not* to be taken as part of the 'mental apparatus', developed by the later Freud, and *not* as an unconscious structure that is defined, according to Hartmann, by its functions).

Freud (1914) recognized more than a century ago that narcissism and the regulation of self-regard are at the service of self-preservation, an insight of fundamental importance for social psychology and personality theory, yet the line of theoretical development through Adler, Federn, Horney, and Sandler to Kohut is a sparsely connected and underappreciated one. Self-regard or self-esteem, being regulated by 'narcissistic object choice' (Freud, 1914) (the use of objects as selfobjects, i.e. for narcissistic purposes) and by behavior strategies aimed at enhancing one's worth and approvability in the eyes of others, refers to one's "confident conviction of being lovable" (Storr, 1968, p. 77), one's implicit knowledge of being acceptable to others and safely embedded in the social milieu. What this means is that one is protected against the *aggressive potentialities* of others. The need for approval and recognition (Flugel, 1945), for the purpose of upholding self-esteem, is equivalent to the striving for coherence of the self (Kohut) and the need to maintain the feeling of safety (Sandler), all of which can be regarded as direct expressions of our evolutionarily ancient need for protection against intraspecific aggression (Konrad Lorenz), against the risk of victimization, expulsion, and annihilation by our fellow human beings (whereby 'paranoid anxiety' [Melanie Klein] is the awareness of this risk). Protection against intraspecific aggression is principally achieved by appeasement or subordination of others and by binding them into a mutually aggression-inhibiting context.

Safety is also felt when narcissistic supplies are received or readily available. Developmentally, the first context within which safety is experienced is the mother-infant relationship (the primary narcissistic fusion with the mother). Self-esteem is similarly based on the infant's earliest experience of his mother, namely the experience of receiving "sufficient loving care" (Storr, 1968, p. 77). The mother-infant relationship is not only the first aggression-inhibiting context but also the template for all later relationships (as appreciated by psychoanalysis in general). It is from the context of 'true parental care' (involving the feeding and grooming of offspring in exchange for infantile care-seeking behaviors) (Eibl-Eibesfeldt, 1970) that various behavior patterns evolved that served the inhibition of intraspecific aggression in increasingly complex social formations.

Humans are, first and foremost, object-seeking (rather than pleasure-seeking) beings, as emphasized by Fairbairn (1952). The primary aim of the person is not libidinal pleasure, as Freud had proposed and early psychoanalysts had maintained, but to establish satisfactory relationships with objects, relationships that provide and recreate the context of security. Object-relations theory emphasizes our dependence on objects (Klein, 1940, 1946; Faribairn, 1952). Self psychology elucidates the nature of this dependence, attributing to objects 'selfobject' functions, that is, the ability to act as sources of narcissistic supplies (approval, recognition, acceptance), thereby maintaining the individual's narcissistic balance (self-esteem, integrity of the self) (Kohut). It is important to emphasize that selfobjects are merely objects (significant others), but through them the self is constituted and maintained in its cohesiveness (by way of mirroring). Joffe and Sandler (1965) formulated this insight thus: the object is "a vehicle for the attainment of the ideal state of well-being" (safety), it "is ultimately the means whereby a desired state of the self may be attained" (p. 158). Wellbeing or safety results from social recognition and approval, that is, from narcissistic supplies or their availability (Joffe & Sandler, 1968, p. 231). The feeling of safety is the developmental extension of the infant's "awareness of being protected ... by the reassuring presence of the mother"; it "develops from an integral part of primary narcissistic experience" (Sandler, 1960a, p. 4). Primary narcissism, as implicated in the earliest relationship between mother and infant, gives rise to secondary narcissism, that is, the regulation of self-regard by relating to (external or internal) objects (Freud, 1914). Primary narcissism was suggested by Sandler and Sandler (1978) to be the origin of the sense of safety or wellbeing, which the individual attempts to regain throughout life by way of relating to objects. It is the developmental departure from primary narcissism that gives rise to ongoing efforts, throughout life, to reexperience feelings of safety by relating to objects *in a way that recapitulates* aspects of the early and earliest relationship with the mother (Sandler & Sandler, 1978). The need for approval from those about us, "for the feeling that we are accepted by society", is "a continuation into adolescent and adult life of the young child's need for the approval of his parents, while the anxiety and despondency caused by the sense of being outcasts from society corresponds similarly to the infant's distress at losing their love and support" (Flugel, 1945, pp. 55-56).

Klein's concept of 'depressive anxiety' (a feature of the 'depressive position' of infantile development) refers to the infant's insight into his dependence, for survival, on the maternal object (and, later, the adult's dependence on a derivative of the primary object). By contrast, anxiety associated with the 'paranoid-schizoid position' of infantile development (*reemerging* later in life as a result of failure in early life to 'repair' internal objects, on whom the infant and later the adult depends) (Klein, 1940, 1946) relates to the potential of aggression from (persecution by) conspecifics and consequential annihilation (including the possibility of aggression from the mother). Lack of close relationships in early life (and failure in childhood to form secure internal objects, equivalent to failure to form a secure self) renders the individual liable to regress to the paranoid-schizoid position, in which fears of persecution and annihilation are reawakened and confirmed (Klein, 1940). Insecurity (lack of securely established internal objects) brings back to the surface paranoid anxiety and the need to monitor suspiciously the world of external objects (Klein, 1940). The danger to which primitive humans would have been exposed early on in hominoid evolution was that of persecution and annihilation by the primal group; and it is to deal with this possibility and in defense against this fear that we have to draw on securely established internal objects and activate selfobject functions of external objects, objects that ensure our self-preservation by supplying us with narcissistic nourishment or having these supplies available for us (and thereby signaling to us that their aggression is inhibited). Developmentally, the role of the mother is taken over by the leader of the group; the internal representative of the mother (the superego) is projected onto the leader by each member of the group. Not only the leader, the group as a whole relates to the individual member in much the same way as the mother relates to the infant (Scheidlinger, 1964, 1968); and the individual's fears of the group and need for protection from the group (dependence on the group) mirror the infant's basic attitudes toward the mother, as described by Klein.

Anxiety arises "out of loss of narcissistic supplies" (p. 136), implying loss of *connectedness* to others and "loss of help and protection" (Fenichel, 1946, p. 44). Anxiety, "the most extreme degree of which is a feeling of annihilation" (p. 134), "means also a loss of self-esteem" (Fenichel, 1946, p. 44). Anxiety is the 'polar opposite' of the feeling of safety (Sandler) implicit in one's connectedness to and acceptance by the group or leader. Anxiety is an awareness of the basic hostility of the group and of the danger of being attacked. 'Basic anxiety' ('basic insecurity') is "a feeling of helplessness toward a potentially hostile world" (pp. 74-75), "a basic feeling of helplessness toward a world conceived as potentially dangerous" (Horney, 1939, p. 173). In a state of basic anxiety, the environment is felt as a menace, "the environment is dreaded as a whole" (Horney, 1939, p. 75). Basic anxiety is a feeling of "impending punishment, retaliation, desertion" (Horney, 1937, p. 235). The danger for the individual consists, in part, in the possibility of being *obliterated* (Horney, 1939, p. 75), that is, being annihilated by conspecifics or the group as a whole. Basic anxiety, arising when "one feels fundamentally helpless toward a world which is invariably menacing and hostile" (Horney, 1937, p. 106), motivates the pursuit of reassurance, approval, and love (i.e., narcissistic sustenance). Receiving others' reassurance, approval, or affection serves "as a powerful protection against anxiety" (p. 96). In soliciting others' approval or affection, we inhibit their innate hostility toward us and counteract our sense of being helplessly exposed to a menacing world. Horney (1937) spoke of "the dilemma of feeling at once basically hostile toward people and nevertheless wanting their affection" (p. 111), a dilemma that is experienced most vividly by neurotic persons as well as patients with schizophrenia (Laing, 1960).

Wilhelm Reich (1928, 1929) was perhaps the first to articulate that a person's character is a 'narcissistic protection mechanism', a mechanism that protects against dangers emanating from an *inherently* dangerous outer world. Indeed, the seeking of a position of safety, a position wherein others' acceptance, approval, or love are forthcoming or available, is the operating principle of the personality. There are different strategies, featuring in different personality types, of recreating the infant's experience of being in the focus of the mother's loving and caring attention, of recreating a state in which acceptance by the mother was felt to be unwavering and unquestionable. Narcissistic needs, arising once the infant recognizes his separateness from the mother (and enters the stage of secondary narcissism), "compel the child to ask for affection", whereby the child may solicit and procure essential narcissistic supplies by way of exhibitionistic behaviors or "by force"; or he may seek to attain them "by submissiveness and demonstration of suffering" (Fenichel, 1946, p. 41). There is, throughout life, a striving to reenact aspects of early and the earliest object relationships, so that "a great deal of life is involved in the concealed repetition of early object relationships" and reenactment of relationship patterns that have from the first years of life operated as 'safety-giving or anxiety-reducing maneuvers' (Sandler & Sandler, 1978). Throughout life, the individual is disposed to employ one or another mode of generating safety, submission being one them, control another, exhibitionism yet another. The aim of predominantly exhibitionistic patterns of relating, not just to another individual but also to the group or an organization, is to *display* an approvable self and to thereby attract narcissistic sustenance (positive attention in the form of approval). Submissiveness and forceful control are methods of generating and maintaining a *context* in which care-giving (narcissistically nourishing) signals can be received from derivatives of the maternal object; they are methods of controlling the *responsiveness* and availability of such derivatives. Personalities differ with regard to the extent to which these methods are woven into their habitual patterns of social behavior.

Through the exercise of power over others, generally involving a sublimated or neutralized form of aggression, the 'purpose of the self' (Horney), which is to maintain or establish connectedness (to the social surround) and thereby to minimize basic anxiety, can be served. Control over the other may involve the threat of abandonment. One induces fear of abandonment in an other, so that one does not have to face abandonment oneself. Making oneself *indispensible* to a common pursuit or an organization (on which the safety of each member depends) is a related method of attaining a position of safety. Compliance, being a derivative of evolutionarily older submissive behavior employed in agonistic encounters (with conspecifics), inhibits intraspecific aggression and thereby generates a context of safety, the context in which the self can express its needs for affection and playfulness. Developmentally, compliance emerges in the mother-infant context for the purpose of upholding inhibition of maternal aggression. Noncompliance disinhibits maternal aggression and, later in development, that of the superego or of external representatives of the superego. Compliance with internal (superego) and external standards flows into many modes of relating to the social surround on various levels, modes that involve appeasement of the superego or superego projections so as to enable the solicitation of narcissistic nourishment from them. The display of helplessness is another strategy for overcoming anxiety and strengthening the self. Basic anxiety "concurs with a feeling of intrinsic weakness of the self"; and this weakness gives rise to "a desire to put all responsibility upon others, to be protected and taken care of" (Horney, 1937, p. 96). The example of 'regression' to a position of helplessness and greater dependency also illustrates the principle that safety-seeking modes of behavior become stabilized in particular environmental or cultural contexts. Not just regression, every mode of social behavior is about recreating conditions under which the mother's care and love were reliably available, whereby the attainment of a position of safety in this way can occur on different levels of social complexity, importantly with greater or lesser reference to the wider social and cultural context. Horney saw in basic anxiety a powerful motivator for social behavior and organizer of the personality, but she did not fully appreciate the fact that patterns of social behavior are in essence patterns of unconsciously relating to the mother and seeking the safety inherent in the earliest relationship. Horney (1937) discerned "four principle ways in which a person tries to protect himself against the basic anxiety: affection, submissiveness, power, withdrawal" (p. 96). These four principle ways, trough which basic anxiety is kept at a minimum, lie at the heart of different types of personality structure. In the neurotic personality, these "moves toward, against, and away from others became compulsive" (Horney, 1950, p. 366).

The superego is an introjected source of approval and disapproval, and as such would "take over the functions of parents or other moral authorities", but "we can never – at any rate within the range of normal mental life – become entirely independent of the approval or disapproval of our social environment" (Flugel, 1945, p. 55). Narcissistic nourishment, in the form of approval or praise, can be attained from the superego or from external superego projects on condition of compliance. Attainment of approval or praise from the superego (from internally imagined or externally projected versions of the superego) involves preparation for or performance of culturally defined social acts, including cultural and religious rituals, and aspiration to or fulfillment of valued social roles, so that both exhibitionism and compliance are brought to bear. The superego is readily projected onto external authority figures or adopts the form of internal images of significant others. God and distant ideological leaders are the clearest examples of superego projections into the realm of imagery. God provides the most striking evidence for the existence of the superego. Religious and other cultural processes in society are founded upon an unceasing and sometimes increasing need to reexperience the safety that was once provided by the mother, whereby both the need for safety and the enduring role of the mother remain unconscious. As counterpart to the externally projected superego, there is the experiential self or 'ego' (in Freud's original sense of the term, and in the sense Federn continued to use it), which captures the feedback (mirroring responses) we receive for our displays of compliance and for our situationally appropriate exhibitionistic or ambitious actions. There is also the self that features briefly and indistinctly in our imagination, which encapsulates our *expectation* of narcissistic sustenance from the social milieu at large (being an abstract superego projection). The latter self is more closely related to (or a manifestation of) the ego ideal (and also related to the ideal self [Sandler] or idealized self-image [Horney]). This imaginary self incentivizes goal-directed behavior; it aids our reality-oriented striving for acceptance by one of the developmental derivatives of the mother (or by a projection of the superego), to be accepted and be thus eligible to receive care and affection. The imagined self can however also be employed defensively in states of detachment. We may be drawn to states of introspection, states in which we imagine our self and the world as it relates to us (to our self) and in which we bolster our self in order to overcome paranoid anxiety, the fear of being deprived of others' recognition and acceptance and being expelled from the group or annihilated. To the extent that social roles have become imprecise and fluid and relationships have become fragile, the self has to be shaped or defined internally for the purpose of pleasing the superego, which then operates as a substitute for a stable external point of reference. We inspect and shape our self for one purpose, that of becoming acceptable to the superego or one of its projections. Neither the experiential self nor the self-image (related to 'ego ideal') exists in itself. The self is always bound (in a dipole) to the superego, to an external derivative of the primary object, or to one or another group; the self relates to (and is structured by) the superego or a projection of the superego.

Self psychology suggests that a stable representation of the self signifies stable connectedness to the selfobject milieu; and it entails a sense of worthwhileness, that is, an expectation that approving or comforting responses will be forthcoming, either from the selfobject milieu itself or from internal self-esteem-regulating structures (essentially the superego). The self of the child is, at first, precariously established and "depends for the maintenance of its cohesion on the near-perfect empathic responses of the self-object" (Kohut, 1977, p. 91). The child phase-appropriately "demands perfect empathy" and "total control over the self-object's responses" (p. 91). A faulty, nonempathic response of the selfobject causes the child to respond with anxiety or rage. 'Optimal frustrations' compel the child to internalize aspects of his selfobjects (Kohut, 1971, 1977). In a process called 'transmuting internalization', narcissistic expectations are withdrawn from selfobjects and transferred to inner structures that perform mirroring (soothing and comforting) functions for the self (Kohut, 1971, 1977); capacities that develop for empathetic self-observation and self-understanding help the child to maintain self-cohesion and self-esteem at times of unresponsiveness of selfobjects (Stolorow, 1983). Nevertheless, our need for selfobjects is enduring. The child's "archaic needs for the responses of archaic selfobjects" (p. 77), for perfect mirroring and merger responses, develops into an "empathic intuneness between self and selfobject on mature adult levels" (p. 66) and an "ability to identify and seek out appropriate selfobjects" that present themselves in the person's 'realistic surroundings' (Kohut, 1984, p. 77). Throughout life, we seek out available mature selfobjects in our social surround, in order to establish empathically resonant relationships with them. Although our selfobject experiences mature, "the archaic selfobject continues to exist in the depth of our psyche; it reverberates as an experiential undertone every time we feel sustained by the wholesome effect of a mature selfobject" (Kohut, 1983, p. 398). This archaic selfobject in the depth of our psyche is nothing other than the superego.

We depart from the assumption that the external world, as we perceive it, is in a fundamental sense equivalent to the consciously experienced inner world. We shall regard the superego as a conscious phenomenon that belongs to the realm of imagery, existing on the margins of consciousness *when the social world is thought about.* Our conscious experience of the *external* social world is, to a substantial extent, an external version, or 'projection' of the superego. The most varying social configurations, including the cohesive group, represent external replicas of the superego and thus of the primary object. We will not in this book focus

on the superego as an unconscious structure, although both
the superego (as an aspect of imaginary consciousness) and
the features and composition of the external social world
(structured in part as a projection of the superego) can be
readily regarded as instantiations of an unconscious
representation that can also be called 'superego'. Likewise,
we shall reassert the equivalence of ego and self, regarding
both as phenomena not only taking shape in imagery but
importantly existing on the margins of the consciously
experienced external social world (whilst acknowledging that
there would be an unconscious *representation* that supports
such self-experience). This will allow us, through our
discussions of psychic processes, to arrive at a relatively
simple model of the personality. The experiential self or ego
is the distillate of simultaneously experienced aspects of the
external social world that relate to oneself. The experiential
self, belonging to social 'reality', would correspond to the
imaginary self, found in the realm of fantasy. This imaginary
self is the ego ideal and serves anticipatory (and guiding)
functions. At the end of the book, we will realize the full
benefit of treating the ego ideal as a form of self-imagery, as
the imaginary equivalent of the experiential self. Sandler et
al. (1963), in view of the widely accepted unconscious
conceptualization of the ego ideal, felt the need to introduce
the term 'ideal self', as the conscious equivalent of the ego
ideal (much as Hartmann had felt the need to define a
conscious self in contradistinction to the unconscious ego).
While Sandler et al. (1963) discriminated between ego ideal
and ideal self, we shall treat ego ideal and ideal self here
synonymously, consistent with the equation of ego with self
(and the return to the spirit of the earlier Freud and the views
of Federn thereafter) (while not denying that there will be
neurally embedded representations that give rise to one or
another set of conscious phenomena, either in externalized
consciousness or in imagery). If the ego ideal were to be
regarded as a conscious phenomenon (albeit an indistinct
and fleeting one), then the various forms of self-imagery
implicated in psychic processes (by Horney and Adler, in

particular) can be unified with much of what is known about the ego ideal and also can be understood more clearly in their relationship to the ego (self) and superego.

Regarding the structure of the book, insights gained by authorities representing different strands of psychoanalytic thinking will be presented selectively and placed side by side, so as to allow the illustration of common and uniting themes. The aim is not to critically discuss psychoanalytic schools or consider the way in which they have become somewhat fragmented or even insulated from each other. Rather diverse psychoanalytic material is reviewed from a common perspective, that which affords centrality to the principle of self-preservation, thereby bringing into focus core processes in the personality that have long been foreseen but that have been insufficiently emphasized and escaped full appreciation in the mist of terminological and conceptual differences that surrounds psychoanalytic theory at large. In particular, the compatibility of self psychology (Kohut, Stolorow, Wolf, and others) with other branches of psychoanalytic theory and the presence of self-psychological insights in the works of earlier and contemporaneous theoreticians (Adler, Arieti, Bergler, Bion, Erikson, Fairbairn, Fenichel, Flugel, Freud, Greenson, Hartmann, Horney, Kernberg, Klein, Laing, Mahler, Money-Kyrle, Nunberg, Rado, Reich, Redl, Riviere, Rothstein, Sandler, Schecter, Scheidlinger, Schilder, Winnicott, and others) will be shown. The subsequent Chapters, each dealing with a particular safety strategy, conclude with brief Summary sections; and an overall synthesis is offered in the Conclusions. It is perhaps best recommended that the paragraphs and sections in these Chapters are read one at a time and repeatedly, so as to allow them to unfold their effect and convey their message.

Chapter 1

Compliance

Aggression against conspecifics, called intraspecific or offensive aggression, and related expressions of disgust, disdain, and contempt toward others are innate, phylogenetically ritualized behavior patterns that are elicited by certain stimulus constellations (Adler, 1927, p. 217; Fenichel, 1946, p. 139; Lorenz, 1963; Hass, 1968; Storr, 1968; Eibl-Eibesfeldt, 1970). Offensive aggression is often expressed in situations that involve dominance claims or resource disputes, including disputes over territory and 'rights' (Blanchard & Blanchard, 1989). Territoriality, rights, and ranking order define to a large part our identity (and ultimately self-esteem). Storr (1968) thought that "there exists within us an aggressive component which serves to define the territorial boundaries of each individual personality" (p. 77), that is, the boundaries of our identity and self. We respond to challenges to our social position or territorial boundaries with offensive aggression; and we employ the same aggression when we feel our self or identity itself is challenged. Others respond in the same way to our infringements of rules and deviations from the norm, because these infringements and deviations ultimately challenge their self and identity. Offensive aggression evolutionarily has the objective of inducing submission in the challenger, the induction of compliance with norms and rules being a cultural derivative. While others' offensive aggression (in the form of punishment or retaliatory aggression) induces conformity in us, others' nonconformity releases our own aggressive potential, which however has to be curtailed and expressed in socially appropriate manners to avoid ourselves becoming the target of the group's aggression or the object of culturally sanctioned forms of punishment.

More deeply than fearing defeat in an agonistic encounter with a conspecific, which we can usually terminate by displaying submission and appeasement toward the opponent, we are afraid of becoming the target of the group's joint aggression, as it carries with it the threat of our annihilation (as can be seen in the 'expulsion reaction' in primate groups [Hass, 1968]). As long as we are compliant and conduct ourselves in normative ways, we keep this fear of the group in check and out of awareness. Conversely, disorders of social conduct in others induce in us feelings of aversion and rejection (Lorenz, 1973, p. 244). Unfortunately, we have a tendency to feel distaste or even hostility for individuals who are handicapped or unattractive, although most of us restrain ourselves and repress this hostility (Berkowitz, 1989, p. 51) (or we develop a tendency, by way of reaction formation, to show kindliness). We have a tendency to laugh at others' handicap. Laughter, evolutionarily, is a form of joint aggression by the group (Hass, 1968; Eibl-Eibesfeldt, 1970). Some children grow up in constant fear of appearing ridiculous and being laughed at (Adler, 1927). Ridicule "leaves a permanent mark on the psyche of children which resurfaces in the habits and actions of their adult lives" (Adler, 1927, p. 68). We can become quite readily the target of others' ridicule, disgust, contempt, or distrust unless we inhibit these reactions through our compliance and normality (Laing, 1960). Expressions of ridicule, disgust, or contempt directed toward ourselves are powerful aversive stimuli that reveal our deepest anxiety, that related to the threat of annihilation. Instinctive aggression discharged jointly against outsiders and those who are unable to conform would have served an adaptive purpose in the evolution of primates as well as in early cultural evolution. Nonconformity was selected against, as was the inability to inhibit anger and aggression in accordance with norms that pertain to a particular social situation. Money-Kyrle (1961) spoke of "the elimination since the dawn of civilisation of those whose undisciplined aggression rendered them least adapted to it", which "may have reinstated innate inhibitions, or developed in us an innate disposition to acquire them" (p. 41).

The social position or status we occupy essentially demarcates our access to narcissistic supplies in complex social configurations. Successful *induction* of submission in a challenger not only reaffirms our social position and rights (of access to narcissistic supplies) but also means that we receive submissive and appeasing signals from the challenger, signals that are evolutionarily related to approving and praising signals. Either type of signal provides narcissistic nourishment, enhancing our self-esteem and restoring our feeling of safety. Furthermore, the *display* of submissive and appeasement gestures, too, is a way of attaining safety. Appeasement gestures, signaling submission to a victor in an agonistic confrontation, are ritualized escape behaviors; their purpose is to attain safety (Eibl-Eibesfeldt, 1970). Submission to an authority (or to a dominant other in a social dependency relationship [pp. 219, 245]), visually communicated by submissive gestures (and verbally by praises of the other's authority), is ritualized escape behavior, too (Rado, 1956). For instance, smiling (having evolved to be as dissimilar as possible from the 'expressive movement' signaling aggressive intent) powerfully inhibits the other's aggressiveness and is frequently used in greeting gestures and to communicate contact readiness (Hass, 1968; Eibl-Eibesfeldt, 1970). Culturally ritualized forms of appeasement gestures (originally used in agonistic confrontations) include bowing and nodding (Hass, 1968; Eibl-Eibesfeldt, 1970). Sandler (1989) pointed out that "[w]e are dependent to an enormous degree upon others for the minute nods of agreement and approval, for signs that friendliness rather than hostility is present, for safety signs" (p. 81). Attainment of safety by displaying submissive gestures toward another and

witnessing the other's peacefulness or pacification can be an
enduring personality feature, one that is particularly evident
in persons of a submissive and morally masochistic
disposition. Similarly, compliance with social norms, laws,
and traditions or with the demands of those who are in a
position of authority (compliance that is verbally avowed at
suitable social and cultural occasions) protects against
anxiety, because it prevents us from being dislodged from or
deprived of our position or status (and potentially being
made an outcast and having to face joint aggression from the
group or society). Compliance, again, not only has the effect
of inhibiting others' or the group's innate aggressiveness but
also protects our rights of access to narcissistic supplies and
preserves our self-esteem.

1.1 Conditionality of Parental Love

Freud (1930) thought that the child's motivation for yielding
to parental commands and expectations and for foregoing
the satisfaction of his drive impulses is to be found in his
helplessness and dependence on his parents for survival
(preservation of the self). For reasons of self-preservation,
the child fears to lose the love of his parents. The child learns
that what is 'bad' is whatever causes him to be threatened
with the loss of parental love (Freud, 1930). The child's self-
esteem is tied up with parental love. As Fenichel (1946)
remarked, the "child loses self-esteem when he loses love
and attains it when he regains love" from his parents (p. 41).
Children "need supplies of affection so badly that they are
ready to renounce other satisfactions if rewards of affection
are promised or if withdrawal of affection is threatened" (p.
41). "The promise of necessary narcissistic supplies of
affection under the condition of obedience and the threat of
withdrawal of these supplies" makes children educable
(Fenichel, 1946, p. 41). The child has to learn to fulfill
parental demands and meet their expectations in order to
access narcissistic supplies and maintain his self-esteem.
The child's self-esteem is "connected with his capacity to

avoid doing what the parents do not want him to do" as well as "with his capacity to do what his parents want him to do" (Arieti, 1970, p. 21). Fulfillment of parental demands and expectations raises self-esteem precisely insofar as it creates conditions conducive to receiving narcissistic supplies from the parents. Mastery of developmental milestones is another source of parental approval. As the child develops, "the most ordinary, most commonplace steps in development become the subject of expressions of approval and/or disapproval" (Arlow, 1989, p. 151). Every act the child performs "becomes for the child fraught with a sense of judgment, approval, or disapproval" (p. 151). What the mother approves becomes 'good', and what she disapproves becomes 'bad'. Pleasurable (narcissistically gratifying) experiences of having done the 'right thing' and the unpleasure associated with having been 'bad' "become the dynamic background against which later, more highly developed concepts of good and evil are examined and processed" (Arlow, 1989, p. 152).

Self-control of desires and impulses, as it is acquired by the infant, is intended to avoid separation from the mother and keep at bay separation anxiety, the anxiety that developmentally arises when the infant becomes aware of the mother's separate existence (Lebovici, 1989, p. 429). The sacrifice of desire and impulses is "the means of assuring oneself of the persistence of parental love" (p. 424), of narcissistic supplies (Lebovici, 1989). Internalization of parental expectations, demands, and prohibitions means that activation of a *mental representation* of the parent (of 'the adult authority-love object') becomes sufficient to suppress or encourage the child's actions (Arlow, 1989, p. 153). What the child thinks of is the mother's potential affective response to his behavior. Thinking of "the possibility of pain from the threatened loss of the mother's love and the fear of punishment by her" suppresses the wish and inhibits its enactment (Arlow, 1989, p. 153). As the superego forms, the child becomes less dependent on narcissistic supplies from the outside (p. 41); the superego becomes the provider of narcissistic supplies and the regulator of self-esteem (Fenichel, 1946, pp. 105-106). "What was formerly a wish to maintain or reestablish harmonious relations with the important objects now appears as a pursuit of inner harmony" (Arlow, 1989, p. 155), the harmony between ego and superego (representing "a freedom of tension experienced as guilt, fear of punishment, and loss of love" [Arlow, 1989, p. 155]). It is the fundamental dependence of the child on the mother for narcissistic supplies that renders the individual, for the rest of his life, fearful of losing the approbation of his social surround or superego and that makes him work toward establishing and maintaining that approbation, whereby the approval or recognition he obtains from others or from the superego is essentially equivalent to the love of the mother (who has been internalized as the benevolent superego, the unconscious 'omnipotent object' [Bursten, 1973], which is reprojected, throughout life, onto external figures or organizations [Flugel]).

For the child to develop an adequate superego, narcissistic gratification has to be consistently contingent on his renouncing the immediate gratification of aggressive, sexual, or other 'instinctual' desires. Parents 'spoil' their child if they provide him with narcissistic gratification unconditionally. If parents do not force on their child "some restraint on the free expression of egoistic desires", that is, if the child is allowed to express his impulses in inappropriate, antisocial ways without consistently incurring punishment or withdrawal of love, then he "has little motive to develop those restraints on impulse that are embodied in the super-ego" (Flugel, 1945, pp. 192-193). Adler (1938) found that "whenever a mother is too lavish with her affection and makes behaviour, thought, and action, even speech superfluous for the child", then the child is more readily inclined to develop a selfish and exploitative attitude; the child "will continually press always to be the centre of attention" (p. 113). 'Spoilt' children "will develop egoistic, envious, jealous traits to a high degree of intensity" (p. 114); they will not cooperate, expect everything, and give nothing. A spoilt or pampered child will "regard it as her right to suppress other people and always be pampered by them, to take and not to give" (Adler, 1938, p. 114). Pampered children lack perseverance and have a tendency to aggressive outbursts (p. 114); they are resistive to "any change in a situation that gratifies their wishes" (p. 42). Their resistance to any change can be active or passive. Whichever method of resisting change and maintaining conditions conducive to the satisfaction of their craving for affection is successful, it provides "a model which is followed in later years" (p. 43); it becomes part of the 'life style' of the adult (Adler, 1938), that is, part of their personality.

1.2 Subordination to a Leader

Members of the group feel secure in the presence of their leader, however "behind the happy security felt in his presence there is a nagging fear of its loss" (Redl, 1942, p. 24). Members feel anxious if they are not sure of their leader's approval, especially if the leader is of a patriarchal type (Redl, 1942). Members enjoy a sense of safety for as long as they behave in accordance with the code of the patriarchal leader. Insofar as members behave in accordance with the code of the leader ('central person'), their superego can be said to incorporate the superego of the leader (p. 24). To be accepted by the leader is a requirement for happiness (and safety), and members are "ready to pay for it by conscientious output of work" (Redl, 1942, p. 25). Members have to repress or modulate their 'instinctual drives' (aggression, sexuality) if they want to be spared the wrath of the leader and be assured his love and approval (narcissistic supplies). Under certain circumstances, instinctual impulses are readily expressed. For instance, when the group is faced by an external threat (Bion's [1952] basic assumption of 'fight or flight'), individuals are freed, as Freud (1921) remarked, from the repression of their aggressive impulses. For the most part, inside or outside the hierarchical group context, the individual exercises self-restraint, repressing his instinctual drives, for the purpose of maintaining his safety. Self-restraint does not have to depend on the presence of a patriarch or leader. The leader has been replaced, in the context of modern society, by the self-observing and self-critical agency, the superego.

Hypnosis illustrates our innate potential for submissiveness to the group leader (Freud, 1921). Adler (1927) remarked that human beings have a "habit of accepting authority without testing it" and are "capable of such submission that [they] can fall victim to anyone who poses as the possessor of special powers" (p. 64). Hypnosis is based on this very capacity, to show servile obedience. The 'critical faculties' of the hypnotized subject are 'paralyzed' as he "becomes, so to speak, the tool of the hypnotist, an organ functioning at his or her command" (p. 64). Individuals who are highly susceptible to suggestion and hypnosis generally tend to overvalue others' opinions and undervalue their own (p. 65). Some people "consider it an honour to appear submissive"; they can be seen "bending forward in the presence of others, listening carefully to everyone's words, not so much to weigh and consider them, but rather to carry out their commands and to echo and reaffirm their sentiments" (Adler, 1927, p. 206). The submissive person lacks self-assertion, readily complies with the potential wishes of others, and avoids at all cost to arouse others' resentment (Horney, 1937, p. 97). Submissiveness and the 'complying attitude', which "may take the form of not daring to disagree with or to criticize the other person, of showing nothing but devotion, admiration and docility" (pp. 119-120), "serve the purpose of securing reassurance by affection" (Horney, 1937, p. 97). We adopt authoritative ideas with varying degrees of readiness and conviction. In persons with hysterical or compulsive character pathology, there may be a greater "readiness to defer to authoritative opinion, to accept ideas and eventually "believe" them, or rather to think that one believes them" (Shapiro, 2000, p. 41). Authoritative rules "make reflection unnecessary and save energy"; "we follow them simply because it feels wrong or, again, would require inconvenient reflection to do otherwise" (Shapiro, 2000, pp. 51-52). Almost as much can be said about rules and ideas that we think are our own but are really means of complying with the superego. Hypnosis also illustrates the principle of projection of the superego (Flugel, 1945). The hypnotized person is highly suggestible insofar as he has projected his superego, the self-critical faculty, onto the hypnotist. The projection is accompanied by regression to an infantile level of functioning characterized by a high degree of trust in and obedience to a parental figure (Flugel, 1945).

1.3 Superego

The ego submits itself to parental standards, at first, and then to the standards of the superego "because it is rewarded by love and protection" (Nunberg, 1955, p. 146), at first from the parents and then from the superego (the introjected parental objects). In order to secure the parents' love, the "ego must renounce above all its genital and sadistic impulses, in so far as they are directed against the parents" (Rado, 1928, p. 59). Restrictions on the discharge of instinctive impulses, which are demanded at first by the parents and then by superego, are accepted by the ego "because it is compensated by narcissistic gratification" (Nunberg, 1955, p. 147). The child (the ego) complies with parental demands and expectations, including demands for modification or repression of his instinctual wishes, in order to receive narcissistic supplies and feel safe; and – having had to face repeated frustrations in his unceasing need for narcissistic sustenance and therefore having had to 'introject' the source of narcissistic supplies and thus form the superego – he complies with the demands of the superego. The influence that the superego continues to exert over the individual's behavior "is a reflection of the child's dependence on his real parents as a source of narcissistic gain in the earliest years of life" (Sandler, 1960b, p. 41).

The danger of loss of parental love is, as Rado (1928) stated, "sufficient to compel formation of the super-ego" (p. 59). The 'good object', "whose love the ego desires, is introjected and incorporated" and "raised to the position of the super-ego" (Rado, 1928, p. 60). As Nunberg (1955) described the process of superego formation, initially "the impressions of objects are absorbed into the ego", but then they detach themselves from the ego and "amalgamate with each other, and thus form an independent structure, the superego" (p. 141). The superego becomes a "psychic agency differentiated from the ego" (p. 140); the superego "stands apart from the ego" and "takes the ego as an object" (Nunberg, 1955, p. 142). In other words, the superego (like a parent) monitors the self (ego) and judges whether the self is worthy of praise or deserving of punishment (disapproval), that is, whether praise or disapproval are likely to be received from external sources (from external superego replicas, including the leader). The superego, containing 'introjects' of the authority of the parents, becomes the major source of narcissistic gratification and self-esteem (and of the feeling of safety) (Sandler, 1960b). In consequence of normative behavior, "the feeling of being loved is restored by approval of the superego", replicating "the affect experienced by the child when his parents show signs of approval and pleasure at his performance" (Sandler, 1960b, p. 39). Likewise, "what was previously experienced as the threat of parental disapproval becomes guilt"; "and an essential component of this affective state is the drop in self-esteem" (p. 39). Guilt is a "warning signal of impending punishment or loss of love" when, however, no potentially punishing agent is present (and the role of the punishing agent is played instead by the superego) (Sandler, 1960b, p. 38).

Parents place "taboos on many potential sources of joy and satisfaction" and inflict punishment on the child "at the slightest hint that these taboos may be infringed" (Flugel, 1945, p. 76). Given that parents are bound to frustrate their children's instinctive strivings, the arousal of anger and revolt against the frustrating parents is inevitable (p. 77). The child's expression of anger against his parents provokes them into acting in retaliation; "they punish him, and withdraw their help, love and approval" (Flugel, 1945, p. 36). Given the predominance of the need for safety, aggressive impulses against the parents have to be inhibited or displaced. They are redirected and turned against the self, even before the parents are introjected, that is, incorporated in the self in the form of the superego (Flugel, 1945, p. 36). With the establishment of the superego, "the inward recoiling aggression also becomes attached to the super-ego" (p. 36), because the superego, representing the forbidding and punishing parents, "is already endowed with the aggression naturally attributed to them as frustrating agents" (pp. 36-37). There is, in other words, "a collaboration between the aggression from outside (or from the corresponding introjected moral authority) and the recoil of the person's own aggressiveness" (Flugel, 1945, p. 80). Aggression against the parents becomes attached to that part of the superego that "corresponds to the child's picture of the parent[s] as ... harsh, forbidding, terrifying, and punishing being[s]" (as contrasted with the part of the superego that corresponds to the picture of them as loving and protecting [and hence narcissistically gratifying] beings) (p. 76). Thus, "the sadism of these authorities and the sado-masochistic relation in which we stand to them in our external life [are] mirrored in the relation between the super-ego and the ego in our internal life" (p. 38) (wherein the sadism of the superego complements the [moral] masochism of the ego). The superego, having taken over the role of our parents, continues to both love and thwart us. By way of projection of the superego, we "find, or invent, good and bad figures that correspond respectively to what were originally the loving and thwarting aspects of our parents" (Flugel, 1945, p. 60).

Fenichel (1946) emphasized the close relationship between superego and the external world; the superego is "the inner representative of a certain aspect of the external world", namely the sphere of "threat and promise, of punishment and reward" (pp. 106-107). Once the demanding and prohibiting functions of authority figures have been 'introjected', superego functions can be "reprojected, that is, displaced onto newly appearing authority figures" (p. 107). Group formation and the belief in authority, for instance, are based on reprojection of the introjected authority figure onto external persons (Fenichel, 1946). What persists throughout life is our "reliance on others as a source of self-esteem" (Sandler, 1960b, p. 42). We are morally sensitive to our environment, inasmuch as we "are rendered happy by the approval and admiration of those about us" (Flugel, 1945, p. 174). The superego reflects the fact that "persons remain influenced in their behavior and self-esteem not only by what they consider correct themselves but also by the consideration of what others may think" (Fenichel, 1946, p. 107). The superego encapsulates the individual's attitudes toward himself; it is, at the same time, an expression of his expectation of others' approvals or disapprovals of himself, an expectation that is inherently uncertain. In the absence of such uncertainty, if "narcissistic support is available in sufficient quantity" from the group or its leader, "the superego may be completely disregarded, and its functions taken over by the group ideals" (Sandler, 1960b, p. 41).

Although "the super-ego provides us with the power of self-regulation in a moral sense", external moral control "is in some respects easier and involves less strain and a lesser expenditure of energy" than "internal control by the super-ego" (Flugel, 1945, p. 174). This explains why there is "a constant temptation to project the super-ego and to find fresh super-ego representatives in the outer world, provided only we can discover external figures that sufficiently resemble the pattern of our super-ego as it has been formed by early introjections" (pp. 174-175). We are therefore liable "to find in the outer world masters who will guide our conduct and heroes who will exemplify our ideals, thus affording us some relief from the greater effort of self-regulation" (Flugel, 1945, p. 175). A leader, if successful, "does indeed inevitably become to a large extent the object of super-ego projection on the part of his followers" (p. 184). The superego can also be projected upon a group or organization or, indeed, the country or nation to which we belong. Our country or state "stands in a symbolic relationship to the parent-figures of our infancy" ("as is revealed in the very words 'patriotism', 'motherland', 'fatherland'"), so that "[o]ur earlier loyalty and obedience to the parents are ... very easily transferred to the state" (p. 290). Control by standards of the group or state, that is, control by an external representative of our superego, replaces "control of ourselves through our (internal) superego", with the effect of losing the "individual critical power and moral sensibility" (p. 182) of which we may otherwise pride ourselves (Flugel, 1945).

1.4 Conflict and Self-Contempt

Neurotic symptoms, like many defenses, are compromise formations between instinctual wish fulfillment and the need to preserve safety (and avoid anxiety) (Sandler, 1985). They, like many defenses, are solutions to conflicts between the need to express instinctual impulses (according to the 'pleasure principle') and the need for self-preservation (as originally attributed by Freud to the 'ego instinct') (Hendrick, 1958, p. 139). Rather than pointing to a conflict between forbidden instinctual impulses and the fear of punishment, neurosis represents a conflict between forbidden impulses and the fear of losing the love of internalized objects (the superego), the love on which the integrity of the self (ego) and the sense of inner harmony (narcissistic equilibrium) continue to depend. What lies behind neurotic conflict is "the individual's wish to regain, in his mind, the lost or disrupted harmonious relationship to those internalized object representations whose good will and love at one time represented the difference between life and death" (Arlow, 1989, p. 160). A predominantly punitive or persecutory superego, one that embodies harsh conditions under which the child had to access narcissistic supplies, pervasively suppresses instinctual impulses or forces them along maladaptive detours, resulting in the formation of hypertrophic defenses and neurotic symptoms. To afford oneself a more direct (and socially less sanctioned) expression of instinctual impulses, demands for narcissistic sustenance have to be foregone. Loss of love from the superego and consequential depression, unhappiness, or self-disintegration can be anticipated when drive-related behavior is not curtailed. In neurotic persons, the superego allows gratification of inner wishes "in exchange for the bribe of depression, unhappiness and self-damage" (Bergler, 1952, p. 23).

The neurotic person tries to maintain his security (safety) by making demands on himself, by complying with 'inner dictates' (representing a complicated system of 'shoulds and taboos') (Horney, 1950). Inner dictates of the neurotic person are equivalent to moral standards and ideals of the nonneurotic person (demands of the superego), however, where moral standards and ideals have an obligating power, inner dictates of the neurotic person have a coercive character (p. 73). Inner dictates are linked to a need to actualize the 'idealized self' (equivalent to the ego ideal). The more the neurotic person is driven to actualize his idealized self, the more his inner dictates "become the sole motor force moving him, driving him, whipping him into action" (Horney, 1950, p. 84). While, for all of us, "there are quick retributions if we do not measure up to expectations", the neurotic person responds to nonfulfillment of ideals with "anxiety, despair, self-condemnation, and self-destructive impulses" (p. 74). When the person consistently fails to measure up to his inner dictates, he hates and despises himself (Horney, 1950). He despises himself inasmuch as his overly harsh superego despises him, and inasmuch as, in his childhood, his overly harsh parent disapproved and criticized him for failures to live up to the demands and expectations placed upon him.

Self-contempt can be externalized in two principle ways, in that one either comes to despise others, or one comes to feel others look down upon oneself (Horney, 1945). Firstly, a person who feels right and superior will more likely despise others when externalizing his self-contempt (p. 118). Externalized self-hatred "appears either as irritability in general or as a specific irritation directed at the very faults in others that the person hates in himself" (p. 120). At the same time, inner coercion, that is, compliance with the exacting demands of the idealized self-image, can be externalized by "imposing the same standards upon others as those under which the person himself chafes" (Horney, 1945, p. 123). When the person's compliance with his inner dictates is externalized, he imposes his standards upon others and makes "relentless demands as to *their* perfection" (Horney, 1950, p. 78). Externalizing his compliance with inner dictates, the neurotic person tries to make other people to do the 'right thing', and he often does so by force (Horney, 1950, pp. 84-85). Indeed, we enforce norms and laws insofar as these norms and laws are critical to defining our self and actualizing our ideal self, insofar as these norms and laws enshrine our right of access to narcissistic supplies.

Secondly, the excessively compliant person, who recriminates himself for his failure to measure up to his idealized image, tends to feel that others have no use for him and look down upon him (Horney, 1945, p. 118). It is painful to despise oneself for any weakness; but when self-recriminations are externalized and others are seen as despising oneself, "there is always hope of being able to change their attitude" (p. 119). Externalization of self-recriminations can manifest as "an incessant conscious or unconscious fear or expectation that the faults which are intolerable to oneself will infuriate others" (p. 121). Increased compliance is a major consequence of this form of externalization. Alternatively, rage against the self "may appear as intestinal maladies, headaches, fatigue, and so on" (Horney, 1945, pp. 121-122). There is a form of externalization of inner coercion that lessens the pressure to comply. Externalization of inner compulsion can take the form of "hypersensitivity to anything in the outside world that even faintly resembles duress" (Horney, 1945, p. 123). The person can rebel against inner constraints when they are externalized and thereby maintain 'an illusion of freedom'; he does not have to admit that he subjects himself to inner coercion and therefore does not have to admit that fails to measure up to his idealized image (p. 125). He can maintain the illusion that he is in fact his idealized image; and it is his idealized image that holds him together (p. 126), preventing the disintegration of his self. In general, externalization replaces inner conflicts with external conflicts (p. 130) and, at the same time, allows the person to maintain a protective illusion about himself, that is, maintain his idealized image (Horney, 1945).

1.5 Persona and Self-Definition

According to Laing (1960), 'false self' or 'persona' refers to one's identity-for-others, an identity that arises in compliance with expectations of the social world (p. 105). The 'false self' or 'persona', being constructed from compliance with others' expectations, is designed to keep in check one's fear of their aggression. Compliance is associated with inconspicuousness, one's "attempts to merge with the human landscape, to make it as difficult as possible for anyone to see in what way one differs from anyone else" (p. 118); and this has the effect of inhibiting aggression from the group (or its leader). The fear that is implicit in compliance prevents oneself from expressing one's own hatred, the hatred provoked by the danger the self faces (which, if it cannot be expressed, is a self-hatred); although in psychosis this hatred is revealed (Laing, 1960, p. 106). The 'false self', discussed by Winnicott (1960b, 1989), is based on compliance, too, on an imperative need to 'be good'. Through a false self, the individual maintains 'false compliance' with external demands. The false self operates defensively, protecting the individual against threatened annihilation. It is the mother's moral code that sets the infant on a path of developing the false self. An excessive need for compliance with maternal wishes results not only in adoption by the child of a false self but also in the systematic suppression of the 'true self'. The true self, an experience of aliveness and spontaneous comfort, then hides behind the false self (Winnicott, 1960b, 1989).

Conformists "keep busy judging how closely they approximate common features of the surrounding world"; their conscious experience is focused on "what goes on around them and how to fit into it unobtrusively" (Schafer, 1997, p. 28). Conformists are prone to experience embarrassment and shame "whenever they judge that they have lapsed from being acceptably expectable and unremarkable" (p. 29). Extreme conformists "constantly try to put an end to spontaneous, unrehearsed, unscrutinized expressions of feeling" (p. 29). Excessive repression of spontaneous feelings means that emotional difficulties are expressed mostly psychosomatically (p. 30). Extreme conformists are intolerant of the pain of rejection, which "precludes emotional commitment to individualized others and sets severe limitations on the sense of aliveness" (p. 29). Their incapacity to spontaneously reach out to others contributes to their weak, fragmentation-prone sense of self. Extreme conformity is "built over fragmented selves and objects" (Schafer, 1997, p. 29).

On the other hand, maintenance by the conformist of 'iron control' of the self and others betrays his sense of omnipotence (Schafer, 1997, p. 32). This sense of omnipotence compensates the conformist's weak self-esteem. The role of omnipotence is particularly evident in those who show 'negative conformism', that is, extreme individualism. The extreme individualist defines his self "by what it must not be" (p. 32); his "goal is the construction of oppositeness as a steady state" (p. 33). The extreme individualist is so concerned with defining his self by way of oppositeness, originality, imaginativeness, and freedom that his poses impress "as parodies or mere gestures of imaginativeness and freedom" (p. 33), and his "subjectivity will be considered theatrical and shallow" (p. 33). Lacking in self-confidence, the extreme individualist has to maintain his illusions of omnipotence, which he does in part by not allowing himself to respect tradition. Like the extreme conformist, the extreme individualist "is self-defining under severe constraints"; and, like the extreme conformist, he does so in order to counteract the fragmentation of self and objects (p. 33). The extreme conformist has a narcissistic ethos (p. 30); the extreme individualist, too, has a predominantly narcissistic internal ethos (Schafer, 1997, p. 33).

Greenson (1958) described 'screen characters', that is, persons who feel the need to use one or another 'screen identity', an identity that features likeable aspects of the self and that serves to maintain in repression a more painful, insecure self-image, thereby preserving the person's 'psychic equilibrium' (p. 118). These persons "are unduly concerned with their social standing and long to be accepted, popular, and entertaining" (p. 113). They have an identity disorder or disorder of the self-image, in that they readily adopt character traits of a likeable person and often have multiple self-images, displaying markedly different sets "of character traits at work and at home, with their family or with strangers", adding to their "difficulty in maintaining a consistent and integrated self-image" (Greenson, 1958, p. 121). Like persons with a narcissistic personality disorder, those with a screen character have a perpetual hunger to find new objects that help them to deny their insecurity and satisfy their narcissistic and libidinal needs, whereby objects are often chosen on a 'narcissistic basis' (p. 131). Like patients with borderline personality disorder, they tend to split objects into good and bad ones, apparently lacking "the capacity to fuse the loved and hated object into a single object" (Greenson, 1958, p. 124).

1.6 Perfectionism and Obsessionality

The façade that a person presents to those around him coincides "with what is regarded as "good"", whereas "what is repressed on its behalf will mostly coincide with what is regarded as "bad" or "inferior"'" (Horney, 1939, p. 228). Whatever does not fit into the person's façade, that is, whatever is in conflict with his style of striving for safety ('neurotic trend'), tends to be repressed. The perfectionistic person pursues the need to appear perfect, to have an infallibly acceptable façade, despite painful consequences (p. 229). Traits that are morally objectionable and therefore do not serve the person's 'glory' (i.e., traits that would undermine his imperative to maintain his self-esteem), such as greediness, plain disregard for others, and vindictiveness, must be covered up or are repressed (or are simply denied) (p. 96); however, even legitimate wishes and spontaneous feelings have to be repressed, if they "render it impossible for him to maintain that façade", if "they would endanger the façade" (Horney, 1939, p. 229). The perfectionistic person's "strivings for perfection lack genuineness"; his "pursuit of moral goals is too formalistic" and has a hypocritical character (p. 213). For him, moral standards "mean nothing more than keeping up the appearance of morality" (p. 230). Perfectionistic trends (overconformity and pseudoadaptation) allow the neurotic person to avoid manifest conflict with others (p. 252); overconformity with social standards and expectations puts him "beyond reproach and attack" (p. 220). Perfectionistic trends arise if the child feels that the world, on which he is dependent (i.e., the parents), is a potentially hostile one (p. 252). The child, feeling uncertain of his acceptability, becomes excessively dependent on the opinion of others (p. 92). The growth of his spontaneous individual self ('true self') is stunted (p. 218), inasmuch as he realizes that "in order to be liked or accepted he must be as others expect him to be" (Horney, 1939, p. 91).

Perfectionistic persons are dependent on others' opinions about them, and "their feelings, thoughts and actions are determined by what they feel is expected of them" (Horney, 1939, p. 215). While the perfectionistic person's "security rests on an automatic conformance with what he believes others expect of him", "he is anxious to hide from himself the fact and the extent of his dependency" (pp. 250-251). Although "we are all dependent on the regard others have for us" (p. 217), and although "everyone living in an organized community must keep up appearances" (p. 216), the person with perfectionistic trends "turns altogether into a façade" (p. 217). For him, all that matters is to measure up to expectations and standards and to fulfill his duties (p. 217), "to maintain the *appearance* of perfection" (p. 215) (whereby this compulsion to appear perfect pertains "to whatever is valued in a given culture" [p. 217]). The Freudian superego can be regarded, according to Horney (1939), "as a safety-device, that is, as a neurotic trend toward perfectionism" (p. 77). The superego, thus conceptualized, is the more or less stringent need to appear perfect, to keep up appearances of perfection (p. 216). Persons who "adhere to particularly rigid and high moral standards" and who have "a passionate drive toward rectitude and perfection" can be said to have a strong superego (Horney, 1939, p. 207). Perfectionism, that is, an excessive compliance with external demands coupled with an overreliance on an approvable façade ('false self'), is a pathological manifestation of narcissism. By conducting himself in accordance with his 'false self', the narcissistic individual lives up to expectations of parental introjects (superego) and aims to attain the approval and love of derivatives of his primary objects. The narcissistic person does not allow himself to really live his own life, to act in accordance with his 'true self', as his only concern is to be accepted and loved by his objects (Miller, 1979).

Greenson (1973) linked the neurotic striving for perfection with a longing to fuse with a perfect (omnipotent) object. The quest for perfection arises in persons who carry "inside themselves, from childhood on, the feeling of not having been loved sufficiently" (p. 487). They are "haunted by the dread of being found unlovable" (p. 487). They "hope for some special form of approval, acceptance" from an idealized, constantly loving mother, the "perfect mother who never existed in real life" (p. 487); and, by attaining perfection themselves, they seek to return to an emotional state in which they "feel a joyous sense of losing their self-boundaries and flowing into, or merging, and becoming one with another person or being – like God, Fate, or Nature" (Greenson, 1973, p. 485). Maternal love and approval, and approval provided by external and internal developmental derivatives of the mother, ensure self-preservation; they provide the context in which the individual is protected against the danger of annihilation by the group, a danger that is dimly felt in states of shame and humiliation. The anxiety against which the perfectionistic person defends himself is the anxiety that "conjures up the danger of condemnation, which is as vital a menace to the perfectionistic type as is desertion to the masochistic type" (Horney, 1939, p. 198). If the person's safety rests on measuring up to his particular standards, if it is solely "rooted in his subjection to rules and to what is expected of him" (p. 226), "then to make a mistake provokes the danger of being exposed to ridicule, contempt and humiliation" (p. 225). The anxiety that is awakened by the possibility of making any mistake or recognizing any shortcoming is 'a fear of being unmasked', a "fear of being found out in all his pretenses", a "diffuse fear that one day he will be unmasked as a swindler, that one day the others will detect that he is not really generous or altruistic but is really egocentric and egoistic, or that he is really interested not in his work but only in his own glory", or that one day "his bluff of "knowing it all" would be called" (Horney, 1939, p. 224). This is not merely the Freudian 'fear of the superego', as suggested by Horney (1939), but a more deeply buried 'disintegration anxiety', to use Kohut's term, or a fear of annihilation, to use Fenichel's term.

Anxiety gives rise to "inclinations towards exactness and order and towards the observance of certain rules and rituals" (Klein, 1932, p. 231). The anxious person seeks to overcome doubt and uncertainty and hence anxiety by being over-precise. Klein (1932) thought that "anxiety belonging to the early danger-situations" is closely associated with the beginnings of obsessions and obsessional neuroses (p. 231). Obsessive acts concerned with order and cleanliness are reactions against an anxiety associated with the child's earliest danger situations, that is, danger situations arising from primitive superego pressures. The severe superego featuring in obsessional neurosis "is no other than the unmodified, terrifying super-ego belonging to early stages of the child's development" (Klein, 1932, p. 229). Superego pressures are not terrifying simply because they stand for the possibility of incurring parental punishment, but because they unearth more deeply buried fears of rejection and annihilation. Pervasive anxiety may not be overcome by compliance alone, especially if the context of a securing relationship is lacking or felt to be lacking. In addition to efforts to appease a critical parent or punitive superego (the inner representative of the critical parent), the person, when in the grip of anxiety, may seek refuge in a world of material orderliness and cleanliness, given that disorderliness and uncleanliness are independent sources of anxiety. Obsessive-compulsive persons have a "chronic need to control the environment as well as the self" (Kernberg, 1996, p. 125); and such control may help them "to feel protected against threatening outbreaks of aggressive rebelliousness and chaos in others" (Kernberg, 1992, p. 29). The obsessional person always watches out for danger; he "has to be ready to avoid any danger from the world outside and to parry, like a fencer in a duel, any possible attack from others"; "the dangers he fears are dangers seen from the perspective of pessimism and distrust; the attacks he fears often have to do with blame and rejection" (Schachtel, 1973, p. 45).

The obsessional person is prepared for a battle with others, whereby his "readiness to fight about logical points and his search for the right rules are intensely emotional" (Schachtel, 1973, p. 46). He is ready "to detect any fault or mistake in them; but he must be equally or even more on the alert and watchful about himself, about any fault or mistake in himself" (p. 45); "he has to be irreproachable, so that he be spared the painful and repressed possibility of feeling unacceptable" (p. 46). Thus, the obsessional person's fear of attack from others is related to a repressed sense of being unacceptable, of being not worthy of others' approval or love, as reflected in his precarious sense of self. The obsessional person may isolate his sense of precariousness from its source in the interpersonal sphere "and experience it primarily in relation to the world in general, to daily routines, to things, to technicalities of their work, and in such well-known phenomena as excessive preoccupation with orderliness, exactness, and the like" (p. 45). "The real source of the uncertainty, precariousness, and doubt in the obsessive-compulsive's life" is, according to Schachtel (1973), "to be found in his pervasive confusion whether the other person and the world in general are friendly or hostile, accepting and approving or rejecting and blaming" (pp. 45-46).

Allied with the obsessional person's quest for perfection (as a precondition for acceptance by a perfect, omnipotent object) are his feelings of omnipotence and his autarchic fantasies. Obsessional ambivalence, which "plays such a decisive part in obsessional (compulsive) neurosis" (p. 50), is a state of "being drawn toward inner passivity and fighting against it with "autarchic" fantasies" (Bergler, 1949, p. 58). The obsessional neurotic seeks to autarchically preserve his narcissism "[b]y changing the role of the passively victimized into that of the active bearer of a doubt" (which is at least "some "activity"!") (p. 58). Magical thought supports autarchic fantasies. The obsessional neurotic confirms his omnipotence of thought by effecting 'guided miracles' (p. 57). It appears to him that 'external fate' enacts his own omnipotence; his observations provide "proof that he is beloved by fate, and therefore has power over it" (p. 58). He thus attains narcissistic gratification (Bergler, 1949, p. 58). Furthermore, the ego tries to attain independence and break off its moral dependence on the superego by reducing the demands of the superego ad absurdum, that is, "by means of arousing the superego to unjust demands" (p. 54). Compulsive behaviors have to be crude in order to effectively reduce the demands of the superego ad absurdum. The obsessional neurotic not only defends against his overly strict superego but also attains narcissistic gratification "by means of this obsessional check-mating of the superego" (Bergler, 1949, p. 55).

1.7 Summary

The superego becomes the provider of narcissistic supplies and the regulator of self-esteem once the parents, as the original providers of narcissistic sustenance, have been introjected and parental demands and expectations have been internalized. Throughout life, the superego is readily projected outward; and social rules and expectations which we think are our own, or with which we think we comply out of our own free will, are but the standards of our superego, the preconditions for receiving love and approval from the superego and for being kept safe by the superego. The superego (when it resides on the margins of internal imagery) is differentiated from and stands above the ego (the self). It monitors the self and judges whether the self is worthy of approval or deserving of criticism and disapproval, that is, whether praise or punishment are likely to be received from external sources, from external superego replicas (superego projects). Feelings of guilt signal that the ego is likely to be punished with disapproval; guilt is an expectation of punishment from external sources, but, in the absence of a consciously elaborated external source, guilt can be conceptualized as withdrawal of love or punishment from the superego.

Aggressive impulses against frustrating parents, at first, and against more diffuse social constraints, later in life, are inhibited, redirected against the self, and reassigned to the superego, so that the superego, apart from being a loving and narcissistically nourishing internal agency, takes on a punitive role as well. We condemn and disparage ourselves for any failure to live up to social expectations inasmuch as, earlier in life, we were disapproved and criticized by our parents for our failures, and inasmuch as we were threatened by them with the withdrawal of their love. Self-contempt can be externalized and turned into righteousness, if others can be despised or criticized for the very faults or failures for which we disparage ourselves. Externalizing the moral standards set by our superego, we impose the same high standards and expectations on others and react with irritability or frank hostility to their violations of norms and laws (thus reprojecting our self-contempt). Owing to projection of the superego and of our own irritability, we would expect that others will be irritated to the same extent by our norm infractions, with the effect of further strengthening our compliant attitude. Extreme compliance, to the extent of inhibiting all spontaneity and affectionateness, gives the personality a compulsive and artificial cast, which impresses as false self or persona or manifests as perfectionism or obsessionality, in either case representing unconscious efforts to be acceptable to and approved by the superego or external superego projects.

Chapter 2

Exhibitionism and Ambition

Compliance wards off intraspecific aggression and thereby defends against persecutory or paranoid anxiety; but compliance is also a *precondition* for the sourcing of narcissistic sustenance (approval and praise). Exhibitionism, on the other hand, *actively solicits* narcissistic supplies in the form of approval or praise. Exhibitionism is concerned not simply with attracting attention but with attracting *positive*, narcissistically nourishing attention from others. Exhibitionistic behavior seeks to gratify narcissistic needs, given that exhibitionism is "always connected with an increase in self-esteem, anticipated or actually gained through the fact that others look at the subject" (Fenichel, 1946, p. 72). Kohut (1966, 1971, 1977) recognized the exhibitionistic nature of the child's expressions toward his mother, expressions designed to attract the mother's interest or praise. Exhibitionism and its developmental derivative, ambition, play an important role in the maintenance of narcissistic equilibrium (homeostasis). The child's exhibitionism, much as the adult's ambitiousness, expresses the need to be seen, to be acknowledged and accepted, and thus to feel safe (and, from evolutionary perspective, to be safe). Feelings of safety (Sandler, 1960a) and self-esteem are manifestation of restored narcissistic equilibrium. Self-esteem is maintained via the creation (investment of 'narcissistic libido' in) and manipulation of 'selfobjects' (Kohut, 1971). Selfobjects are first and foremost objects (significant others). The term 'selfobject' emphasizes the significance of objects (usually other persons, but also groups and organizations and even material objects) for sustaining the self and upholding self-esteem. Selfobjects are objects that are used narcissistically; they provide mirroring and approving responses or allow merger experiences. Selfobject functions of objects (significant others) can be activated by way of exhibitionistic displays or by their idealization. The child's primitive narcissism, expressed through idealizations and exhibitionistic displays, matures into "realistic self-esteem, the ability to be guided and sustained by realizable ideals", and the self's "ability to seek and find realistically available other selves who will sustain it by functioning as mirrors and ideals" (Kohut, 1983, p. 396).

Children age-appropriately perform actions of 'showing off' or clowning in order to attract others' attention (Reddy, 2003). Attention is sought, by way of exhibitionistic behavior, in order to maintain the cohesiveness of the self. Exhibitionistic actions, such as clowning or joking, can be performed in the service of defense, as defensive maneuvers directed at *restoring* self-esteem (Sandler, 1985, p. 105). Defensive exhibitionism in adults is linked with disturbances of narcissism (Sandler, 1985, p. 105). Exhibitionism, when used defensively, is "a technique for gaining admiration and praise in order to do away with underlying feelings of unworthiness, inadequacy, or guilt" (Joffe & Sandler, 1967, p. 181). Exhibitionistic actions, borne out of insecurity and a heighted need for narcissistic sustenance, often have the opposite effect, attracting negative attention and inadvertently increasing anxiety. They attract attention nevertheless, which makes it clearer that exhibitionism is a developmental or evolutionary elaboration of proximity-seeking behavior and separation cries associated with separation anxiety. Exhibitionism is often not evident as such; in sublimated forms, it underpins a great variety of adaptive social behaviors. Exhibitionistic means of attracting others' positive attention (and thereby upholding the narcissistic homeostasis) have to be sublimated, that is, they have to take into account external 'reality'. Actors, for instance, attempt to sublimate their exhibitionistic impulses in their work (Joffe & Sandler, 1967, p. 181), whereby sublimation, an adaptive ego activity (Hartmann), ensures that the safety-enhancing effect of being in the focus of others' attention is not offset by greater vulnerability to others' aggression (and ridicule), a vulnerability that is signaled to the ego (self) by feelings of paranoid anxiety (and shame). Thus, schizoid individuals may engage in artistic activities as a way of expressing their exhibitionism without exposing themselves directly to potentially dangerous social contact (Fairbairn, 1952).

According to self-psychology, "the guiding force in human development is the need for connections to ... self-objects" (p. 102); that is, selfobject experiences and the self are "the primary organizers of psychological development" (Tolpin, 1986, p. 101). The child's 'nuclear self' comprises the 'grandiose-exhibitionistic self' (the seat of 'grandiose-exhibitionistic urges') and the 'idealizing self', formations that come into being as the child tries to regain the sense of omnipotence and perfection that pervaded the primary narcissistic state (Kohut, 1966, 1971). The grandiose-exhibitionistic self looks for the caregiver's gleam and smile ('the gleam in the mother's eye' [Kohut, 1966]), while the idealizing self wants to be picked up and calmed by the caregiver. The early maternal selfobject fulfils a mirroring function (satisfying the child's desire to be looked at and admired), whereas the early paternal selfobject usually serves as an idealizable target. In serving these functions, parental selfobjects enhance the child's self-esteem and support the development of a cohesive self (Kohut, 1983). Self-psychology posits that "the development of the self in relation to its selfobjects" provides "the supraordinate framework" for "the traditional drive-conflict-structural model of the mind" (Tolpin, 1986, p. 102). In the healthy personality, the narcissistic use of objects (as selfobjects) is coupled with an ability and readiness to make libidinal ('selfless') investments in these objects. Kohut (1971) distinguished selfobjects from 'mature objects'. Mature objects are cathected with 'object libido', while also being cathected narcissistically. Mature objects respond to the subject's care, affection, or love with reciprocal care, affection, or love and thereby enhance or maintain the subject's self-esteem (Kohut, 1971). Libidinal investment in objects (establishment of love relationships) thus provides a complementary way of securing narcissistic supplies, one that lowers the individual's reliance on exhibitionistic techniques. Fenichel (1946) stated that the capacity for object love makes available a higher, postnarcissistic source of self-esteem (p. 85).

The 'idealized parent imago' is a psychic configuration that is established by the child relating to a selfobject that is the target of his idealizations, the object of his 'idealizing self'. The idealized parent imago establishes itself (in the child's psyche) if the selfobject responds empathically and with enjoyment to the child's idealizations (Kohut, 1971, 1977). As the self matures, the idealized parent imago is transformed into a structure that supplies healthy ideals ('ego ideal' [Kohut, 1966]). Kohut (1977) proposed that the cohesive self – which develops as the child's narcissistic (mirroring and idealizing) needs are met and 'optimally frustrated' – has two poles, namely the 'pole of ideals' and the 'pole of ambitions'. The pole of guiding ideals is consolidated in the relationship with the admired and idealized, usually paternal selfobject. The pole of ambitions, which crystallizes out of the approving and mirroring relationship with the usually maternal selfobject, is responsible for strivings for power and success. The adult's healthy ambitions are thus an expression of narcissistic needs and a 'tamed' variant of the child's exhibitionism and grandiosity (Kohut, 1971, 1977). Enjoyment of success signifies satisfaction of grandiose-exhibitionistic tendencies, of urges that were originally concerned with inducing maternal mirroring responses. If grandiose-exhibitionistic urges are thwarted, shame arises. The ego ideal, which encapsulates parental expectations and wider social and cultural standards, helps to ensure that exhibitionistic urges (and expressions of grandiosity) are not thwarted; the ego ideal offsets narcissistic vulnerability and offers protection against shame (Kohut, 1966). In terms of the 'bipolar self', the pole of ideals guides ambitions (the derivative of exhibitionistic impulses) into realistic channels; the individual, in Kohut's formula, is *driven by his ambitions while being led by his ideals*. The two poles, together with an intermediate area of 'skills and talents', constitute a 'tension arc' that propels the individual to realistic and constructive action (Kohut, 1977).

Self-esteem, as already pointed out, has its origin in infantile narcissism and omnipotence (Freud, 1914). Infantile narcissism is shaped and channeled into realistic dimensions by the encouragement the child receives from his social milieu for his functional achievements (Erikson, 1959, p. 191). Erikson (1959) saw "the many steps in child development which, through the coincidence of physical mastery and cultural meaning, of functional pleasure and social recognition, contributes to a more realistic self-esteem" (p. 194). Tangible social recognition is also awarded for the acquisition of skills and knowledge during the latency period (p. 201). What develops in this process is "a defined ego within a social reality", a sense of 'ego identity' (p. 194); whereby "the self-esteem attached to the ego identity is based on the rudiments of skills and social techniques which assure a gradual coincidence of functional pleasure and actual performance, of ego ideal and social role" (Erikson, 1959, p. 199). Skills, talents, and knowledge become the means for defining our ego identity and securing a *valued* social role, and thereby for attracting approval from the leader of the group or, unconsciously, from the superego. Groups are evolutionarily linked to the mother-infant relationship; and patterns of relating in groups are essentially those seen between mother and infant. The leader of the group is a derivative of the primary maternal object and also an external representative of the superego. The leader of the group to which we belong "is *par excellence* the figure on whom our super-ego is projected" (Flugel, 1945, p. 184). Through approximation of the ego ideal (brining to bear our skills, talents, and knowledge), we enhance our approvability in the eyes of the superego and safeguard the goodwill of the superego.

Man evidently has a need for the society of his fellows; and "where this instinct is strongly developed it represents more particularly the need to collect and accumulate a specially large measure of love, support and so security, which will be available as a perpetual reserve to be drawn upon at need" (Riviere, 1937, p. 24). "By collecting goodness all round them, which they can dip into at any moment", individuals with strongly developed narcissistic needs "re-create for themselves (by their unconscious phantasy-attitude) a kind of substitute mother's breast which is always at their disposal and never frustrates or fails them" (Riviere, 1937, p. 24). All of us work to attain prestige, power, and possessions because the attainment of such attributes facilitates access to narcissistic resources in our social environment, resources that replicate in unconscious fantasy the mother's loving attention. *Material* objects, once acquired, enhance our self-esteem and instill in us feelings of enthusiasm (Greenson, 1962). Possession of an idealized material object "brings with it a feeling of being joined with it, a feeling of incorporating it, identifying with it, and making it part of the self" (p. 174). The idealized material object enlarges the self; and when the self-image approximates the ego ideal, temporary fusion of the ego (self) with the superego can occur. This fusion is responsible for feelings of elation. The person who is 'enthused' by virtue of possessing an idealized material object feels grander in himself and closer to God, literally inspired by God (Greenson, 1962, p. 175). In hypomania and mania, there is similarly fusion of the ego with the superego, which is similarly based on aggrandizement of the self, whereby in pathological states such aggrandizement is maintained in *denial* of external reality (Greenson, 1962). The merger of the ego (having approximated the shape of the ego ideal) with the superego replicates the infantile experience of fusion with the mother (Rado, 1928, p. 54) and brings with it 'a sense of power and harmony' (Flugel, 1945, p. 55), that is, the power of liberated 'instinctual energy' and the sense of restored narcissistic equilibrium (Fenichel, 1946, p. 407; Greenson, 1962, p. 180).

2.1 Ego Ideal

Narcissism, as Freud (1914) recognized, urges the subject to respect cultural norms. The ego ideal defines standards which, if the subject lives up to them, allow him to gain narcissistic satisfaction (Freud, 1914), that is, restore his sense of wellbeing or safety (Sandler). The ego ideal demands repression of libidinal 'instinctual' impulses, insofar as they are in conflict with cultural norms (Freud, 1914); but it also guides and controls the expression of exhibitionistic impulses, that is, impulses of the 'narcissistic self' (Kohut, 1966) (grandiose-exhibitionistic self). Developmentally, the child performs commended actions and refrains from undesirable ones not only out of fear of punishment but also in expectation of praise (narcissistic gratification) (Nunberg, 1955). The ego ideal is the shape of the self that is most likely to attract approval and praise from the parent and least likely to provoke the parent's wrath (Nunberg, 1955). As a result of oedipal disappointment in the parent, the parent's 'loving-approving and angry-frustrating aspects' are internalized and "become the approving functions and positive goals of the superego, on the one hand, and its punitive functions and prohibitions, on the other" (Kohut, 1971, pp. 47-48). Therefore, the source of narcissistic gratification (love and approval), having originally been the parent, is now the benevolent aspect of the superego, the introjected primary object. The benevolent superego 'smiles upon the self' (Rothstein, 1979) when the subject acts exhibitionistically within the limits that were prescribed or imposed by the parent, when, in other words, the self acts in guise of the ego ideal. Exhibitionistic impulses are coupled with and guided by grandiose cognitions. The child's 'infantile megalomania', that is, "the child's boasting about what he can do and will do in the future" (p. 13), is an important factor in the formation of the ego ideal (Bergler, 1952). The ego ideal "enshrines all the grandiose ideas the child has built up in speculating about his own glorious future" (Bergler, 1952, p. 257). Much as the child's exhibitionism and expressions of grandiosity had previously elicited the mother's smiling response, so the tamed grandiosity of the self will henceforth attract the smile of the superego (or of its external replicas). The child's grandiosity also reflects his desire to be *like* the parent whom he idealizes and endows with omnipotence; and this desire creates a constant inner demand on his ego, a demand that contributes to the formation of the ego ideal (A. Reich, 1953). The ego ideal is formed, as Freud first argued, by identification with the idealized oedipal object; and it acts as a goal to be attained and a blueprint to be emulated. Accordingly, the Freudian ego ideal corresponds, in Kohut's bipolar model of the self, to the 'pole of ideals', which, too, arises from idealization of the parent and has a goal-setting function (Wallerstein, 1983).

The 'ego' in Freud's earlier writings (including in *On Narcissism* [1914]) refers to the self, however after Freud conceived the ego as part of the 'mental apparatus' (along with id and superego), the need arose to differentiate between the two. While Federn (1952) continued to treat the ego as the self (as is done also in this book), Hartmann (1964) proceeded to define the ego by its 'functions'. It became necessary to state that it is the self (or self-representation), and not the ego, that is narcissistically cathected (Sandler, 1960b; Hartmann, 1964; Jacobson, 1964) or, in other words, maintained in its cohesiveness by external (and internal) approvals (Kohut). Insofar as a conceptual distinction arose between 'self' and 'ego', it was felt necessary to differentiate the ideal self from the ego ideal. Sandler et al. (1963) defined the 'ideal self' as the shape of the self that, under particular circumstances "and under the influence of particular instinctual impulses of the moment", "would provide the highest degree of narcissistic gratification" (p. 85). By adopting the shape of the 'ideal self' (i.e., identifying with the ideal self), the child can obtain narcissistic gratification from his parents (secondary narcissism) and experience feelings of triumph and safety. At the same time, the ideal self is the shape of the self that "would minimize the quantity of aggressive discharge on the self" (p. 85). The ideal self is thus determined not only "by the child's need to gain the love and approval of his parents or introjects" but also by the need to avoid their disapproval (Sandler et al., 1963, p. 85). The ideal self is a compromise formation "between the desired state of instinctual gratification and the need to win the love of, or to avoid punishment from, authority figures, internal or external" (p. 85). If the child or adult fails in his efforts to identify with or attain the shape of his ideal self, "then he will suffer the pangs of disappointment, and the affective states associated with lowered self-esteem" (Sandler et al., 1963, p. 88). Mental pain, reemerging whenever the child experiences discrepancies between actual and ideal states of the self, promotes personality development. In the process of 'individuation', ideal states of the self that have become inappropriate are given up, and 'new phase-specific reality adapted ideals' are acquired (Joffe & Sandler, 1965, p. 174). Regression, accordingly, can be understood as a return to a previous shape of the ideal self (Sandler et al., 1963).

The ideal (or idealized) self-image guides the person's efforts to recapture his sense of safety, the safety that is associated with being loved or accepted and recognized by others, of being considered worthy of their attention and approval. Endeavors to actualize the ideal self are an integral part of the personality (and structure the bipolar self) and are generally socially adaptive. What characterizes the idealized self of the neurotic person is the 'fantastic nature' of self-glorification (Horney, 1950, p. 32). The neurotic person tries to turn himself "into this special kind of perfection prescribed by specific features of his idealized image" (p. 25). Attempting to compensate for an inner insecurity, the neurotic person is subject to a "compulsive drive for worldly

glory through success, power, and triumph" (p. 368). Under
the tyrannical system of inner dictates (the 'shoulds and
taboos'), "he tries to mold himself into a godlike being"
(Horney, 1950, p. 368). It is the intensity of narcissistic need
that renders self-imagery fantastic and its actualization
compulsive and potentially maladaptive. Imagery of the
idealized self can be uncoupled from the striving for self-
actualization; then imagery of the idealized self would "take
the form of imaginary conversations in which others are
impressed or put to shame" (Horney, 1950, p. 33). Neurotic
pride, as contrasted by Horney (1950) with realistic self-
esteem, "rests on the attributes which a person arrogates to
himself in his imagination, on all those belonging to his
particular idealized image" (p. 90). Neurotic pride would be
more sustained, if the person's glorified version of himself
could be actualized through accumulation of power,
prestige, or possessions. Work and intellect would be
required to transform the virtues of the idealized image into
assets of which the neurotic person can be *proud* (Horney,
1950). An alternative method of sampling the safety
associated with the idealized image is for the person to
identify himself with it, to convince himself (unconsciously)
that "he *is* his idealized image" (Horney, 1945, p. 98).
Neurotic persons of an arrogant (overtly narcissistic)
character disposition presume, using self-deception rather
than self-actualization, that others *do* regard them as
wonderful (Horney, 1945, p. 112).

The neurotic person is highly vulnerable to disregard,
humiliation, and ridicule. He feels weak "in a world peopled
with enemies ready to cheat, humiliate, enslave, and defeat
him" (Horney, 1945, p. 101). His sense of vulnerability
reflects a weak self-esteem and a deep-seated (and usually
concealed) sense of worthlessness and hopelessness.
Exhibitionistic, self-assertive, and affectionate tendencies
cannot be expressed freely in the presence of strong fears of
rejection or ridicule. Owing to "the ghastly prospect of
ridicule", neurotic persons "do not dare to make themselves

attractive, to try to impress, to seek a better position" and "do not dare to approach people who seem superior to them in any way" (Horney, 1945, p. 152). Neurotic inhibitions, concerning healthy exhibitionistic, self-assertive, and affectionate tendencies, lie at the heart of neurotic disorders (Schultz-Henke, 1951). The idealized image expresses the neurotic person's heightened need for approval and, acting as a 'saving mirage', offers a solution to neurotic conflicts (Horney, 1945). Spontaneous self-expressive tendencies (the tendencies of the 'true self') are inhibited not only out of fear of rejection but also in an attempt to keep alive the idealized image on which the neurotic person depends for preservation of his self and safety. The idealized image excludes everything that is regarded as a shortcoming and that would increase his sense of vulnerability to rejection or humiliation. Not only does the neurotic patient fear failure, disgrace, and ridicule, he is afraid of "all that falls short of glory and perfection"; he "is afraid of not performing as superbly as his exacting shoulds demand, and therefore fears that his pride will be hurt" (Horney, 1950, p. 101). Persons who are "held together merely by their idealized image" exercise excessive *self-control* in an attempt to keep in check impulses that are disruptive to or in conflict with their idealized self-image, whereby "the greatest degree of energy is directed toward the control of rage" (Horney, 1945, p. 137).

2.2 Striving for Superiority

Character, according to Adler (1927), comprises "the tools used by the total personality in acquiring recognition and significance" (p. 135). The striving for recognition and significance can "take many different forms, and every human being approaches the problem of their personal significance in an individual way" (p. 137). The personality strives toward a definite final goal (pp. 68-69), one "that is constantly present, more or less consciously" (p. 229), and that represents an 'ideal state' of being recognized and accepted, a state of security, not only "security from danger, but that further element of safety that guarantees our continued existence under optimum circumstances" (Adler, 1927, p. 32). Adler (1938) understood that, in their present cultural context, "human beings are in a permanent state of feeling their inferiority, which constantly spurs them on to further action in order to attain greater security" (p. 78). This inferiority and its implicit insecurity have their origin in the infant's sense of weakness and helplessness (Adler, 1927, p. 66). The degree of security that human beings demand in relation to the everyday realities is determined early in their lives (Adler, 1927, p. 32). Thus, being guided by a definite goal conjured up in imagination, the personality moves from a state of insignificance and inferiority, through gradual attainment of recognition and significance, to a state of security. Adler (1938) regarded "character as a guiding-line for the goal of superiority" (p. 165). Superiority means eminent or dominant status in the group and power over others; it is thus associated with a measure of security, that is, protection against others' aggression. Superiority also means heightened self-estimation, such as in the form of vanity, which creates more of a deception of security. The striving for superiority or power manifests as ambition or vanity, whereby "a degree of ambition and vanity appears in all human beings, according to their individual method of striving for power" (Adler, 1927, p. 229).

Ambition and vanity and the desire for recognition and attention develop excessively, if the individual, early in his life, felt deprived of love and affection and was unable to express his affection ('social feeling') without fear of rejection (Adler, 1927, pp. 66-68). If the child was not sufficiently allowed (or did not sufficiently allow himself) to experience security through spontaneous and affectionate interactions with his parents, then the solicitation of approval and admiration, being an alternative means of attaining security, becomes firmly enshrined in the personality. The more the individual's 'social feeling' was strangulated in early development, the more preoccupied he tends to be with himself (his self) and with the impression he makes on others; and the more his life is dominated by the goal of superiority, which he strives to attain by exerting power over others or by soliciting their admiration (Adler, 1927, p. 66). Neurotic individuals, in particular, are handicapped by a pervasive sense of insecurity, which they struggle to overcome in search of a sense of great personal worth, a sense of superiority (Adler, 1938). The compulsive striving of neurotic individuals away from inferiority and insecurity toward superiority and security can take the form of self-centered ambition, greed, or envy (p. 121); or it can manifest in "the bearing, character-traits, and thinking of individuals conscious of their own super-human gifts and abilities", in "vanities in connection with personal appearance", in "arrogance, exuberant emotion, snobbism, boastfulness, a tyrannical nature, nagging, a tendency to disparage", "an inclination to fawn upon prominent people or domineer over people who are weak, ill or of diminutive stature", and in tendencies of excessive enthusiasm, "habitually loud laughter" or "habitual excitement over trivial happenings" (Adler, 1938, p. 94).

Ambition is concerned with gaining power, prestige, and possessions. Freud noted in his paper *On Narcissism* (1914) that achievements and possessions serve the function of enhancing self-regard (self-esteem). Privileges and possessions mark out our access to narcissistic resources (to the means of strengthening and protecting our self or ego). They "stand as proofs to us, if we get them, that we are ourselves good, and so are worthy of love, or respect and honour, in return"; and "[t]hey also defend us against our fear of the retaliation, punishment or retribution which may be carried out against us by others" (Riviere, 1937, p. 27). On a deeper level, the striving for power, possessions, and prestige reflects our dependence on the internalized omnipotent object (the benevolent aspect of the superego), which contrasts with dependency behavior in relation to an external replica of the primary maternal object. In either case, the blueprint for safety is the mother's caring and approving attention. All of us need narcissistic supplies and seek to secure our access to these, and thereby to alleviate fears of aggression from others; however narcissistic persons, being incapable of genuine attachment and libidinal involvement, excessively rely on achievements and possessions as a means of regulating their self-regard (Freud, 1914). Narcissistic persons are "absolutely governed by a need to collect reassurances against supposed dangers" (Fenichel, 1946, p. 479). They may do so by collecting approval from their superego or by collecting more tangible narcissistic supplies in the form of affection and confirmation from those around them (Fenichel, 1946, p. 479).

Horney (1937) suggested that the quest for power, prestige, and possessions – deeply embedded in Western culture – is a way of gaining protection against helplessness and against insignificance (protection against 'basic anxiety') (p. 171). Not only infants, humans in general are inherently small and weak. They are weak and vulnerable as individuals, as long as they *are* individuals, as long as their individuality and identity (self) *set them apart* from their group. Erikson (1950) realized that the child's sense of smallness lives on in the adult's latent anxieties, in his sense of vulnerability (pp. 364-368). The adult's achievements and triumphs compensate for this very smallness, overcoming a sense (within him) of danger "from some enemy ... in the outer world" (p. 365) and preventing, at the same time, a "sudden loss of attentive care" (Erikson, 1950, p. 368). Firm embeddedness in the group (with dissolution or dedifferentiation of individuality) is the equivalent of the child's *unconditional* acceptance and protection by the mother. The child develops the self (ego) as an internal structure compensating for lack of firm relatedness to the social surround; and this self forms out of the child feeling that his protection and acceptance by, and entitlement to attentive care from, the mother are *conditional* on his behavior and attitudes. Striving for self-actualization and superiority is continuous with the striving for recognition and acceptance by the group (and its leader); but the striving for self-actualization is the stronger the less the individual was accepted (and loved) by his parents and the less firmly he is embedded in a cohesive group (firm embeddedness in the group being the derivative of an earlier secure mother-infant relationship), the more, that is, he *is* an individual (and the more he sees himself as such).

The striving for superiority, which manifests as striving for power, prestige, or possessions, can also be traced to infantile grandiosity and omnipotence, the child's omnipotent denial of his vulnerability and of his dependence on the mother. Achievements and possessions corroborate (and root in reality) remnants of the primitive sense of omnipotence (Freud, 1914). Adler (1927) thought that the striving for acquisition of 'worldly goods' and the heaping up of possessions allow the individual "to come close to possessing the power of an enchanter" (p. 178); that "individuals who spend their lives chasing after gold are spurred on merely by their vein desire for God-like power" (p. 178), that is, for omnipotence. In the adult's striving for superiority, the desire for acceptance by the mother and the desire for independence from the mother coexist, much as in Kohut's model the longing to be accepted and smiled upon by the mother interacts with (and is in dynamic equilibrium with) infantile grandiosity. The striving for superiority expresses, in other words, both a wish for acceptance by an internalized derivative of a primitive and hence omnipotent object and a striving to deny this very need for acceptance and thus to deny one's dependence on others and more particularly on the group and its leader; whereby the group and its leader, too, are developmental derivatives of the primary maternal object (Scheidlinger, 1964, 1968).

Horney (1937) emphasized that the "striving for power serves as a protection against the danger of feeling or being regarded as insignificant" (p. 166). By exerting power, gaining prestige, or piling up possessions, one protects oneself against basic anxiety. While the normal (and quite ubiquitous) striving for power protects against anxiety (keeping basic anxiety out of awareness), "neurotic striving for power ... is borne out of anxiety ... and feelings of inferiority" (Horney, 1937, p. 163). Possibly for this reason, the neurotic person's 'search for glory', his need to actualize an idealized self-image, is more compulsive than the healthy person's striving for success and 'self-realization' (Horney, 1950, pp. 37-38). The neurotic person's "relentless chase after more prestige, more money, ... keeps going, with hardly any satisfaction or respite" (Horney, 1950, p. 30). Horney, like Kohut, recognized the interaction in the personality between ambition and ideals. The striving for power, prestige, or possessions is guided by an image of oneself occupying a desirable position or playing an influential role, a position or role that is sure to attract others' approval. The capacity of imagination is at the service of both the neurotic search for glory and the healthy striving for self-realization. An ideal to be realized, that is, the desired future shape of the self, has to be conjured up in imagination, however, in the neurotic person, 'checks on imagination' are weak (Horney, 1950, p. 35). The neurotic person "is aversive to checking with evidence when it comes to his particular illusions about himself" (p. 36). He fails "to recognize limitations to what he expects of himself and believes is possible to attain", because the "need to actualize his idealized image is so imperative" (Horney, 1950, p. 36).

2.3 Characterological Defenses

Sublimation of an 'instinctual wish' refers to the *safe* 'gratification' of that instinctual wish. Sublimation means that a drive impulse is expressed in an acceptable, socially appropriate manner. Sublimation is a 'mechanism of defense' insofar as it avoids or minimizes anxiety, the anxiety that is associated with the possibility of disapproval and criticism by others (Laughlin, 1970; Sandler, 1985). Sublimation preserves the feeling of safety; it 'defends' the ego or self, much as inappropriate expression of the drive impulse would endanger the self and, when considering from an evolutionary perspective, threaten the continued existence of the individual. Aggressive impulses are commonly sublimated or 'neutralized' (Hartmann, 1964), but so are exhibitionistic impulses, as illustrated by performances of arts or music (Sandler, 1985). It is precisely inasmuch as the aims, toward which instinctual 'energy' is diverted, are socially acceptable and looked upon by society with favor that these aims are also more acceptable to the ego or self (Laughlin, 1970, p. 312) and hence are pursued by the ego or self. Not only does sublimation or neutralization prevent anxiety and safeguard the ego or self, it also positively strengthens the ego or self, in that it ensures that the individual receives narcissistic supplies in the form of approval and praise from the superego or social surround (an external superego projection). Sublimation strengthens the ego or the integrity of the self and enhances self-esteem by giving the individual access to narcissistic sustenance, the very sustenance that constitutes the self in the first place.

The acquisition of possessions and prestige tends to be accepted and encouraged socially, but if it is thinly veiled and seen as one's sole preoccupation, it invites disapproval. Reaction formation steers against others' perception that one is driven by greed; and it thus protects the self (constituting another mechanism of defense). Greed, like envy, is a ubiquitous but objectionable characteristic; and so it may have to be concealed "behind an exaggerated generosity, which amounts to nothing more than the giving of alms" (Adler, 1927, p. 183). Gestures of generosity, when arising from reaction formation, not only protect the self (against criticism and ridicule); they also "bolster one's self-esteem at the expense of others" (Adler, 1927, p. 183). Reaction formation, in general, "substitutes supposedly desirable and ethical qualities for unrespectable or immoral ones" and, in particular, covers egodystonic greed and selfishness with an exaggerated deference or humility (Flugel, 1945, p. 70). Reaction formation serves the purpose

of the ego ideal, "preventing the ego from falling, or
appearing to fall, too far below the standard required by the
ideal" (Flugel, 1945, p. 69). Reaction formation works
alongside the defense mechanism of repression. In reaction
formation, the ego adopts an attitude that is opposite to what
needs to be *repressed*. Reaction formation can be defined as
the development of a 'pattern of attitudes and reactions' that
suppresses the expression of 'contrary impulses' (Laughlin,
1970, p. 281). Identification aids reaction formation. In
reaction formation, there is "at the same time an identification
with an attitude of the parent or of some other authority
figure", "an attitude which is opposed to the impulse or to the
affect concerned" (Sandler, 1985, p. 102). After introjection of
the parental figure, this becomes an identification of the ego
with the ego ideal.

Through the adoption of "reactive patterns of excessive
kindness, fairness, morality, and idealism, humans seek to
protect themselves against the dangers resulting if they were
to follow their inner impulses of anger, sex, or rage"
(Laughlin, 1970, pp. 287-288). Both, reaction formations and
sublimations, ensure that the person is accepted and
approved in his culture. Both are part of a person's
'characterological armor' (Reich, 1928, 1929). Anna Freud,
too, stated that reaction formations are part of the character
or personality, although she thought that sublimation is
much more an *activity*, an activity that safely expresses an
instinctual impulse (Sandler, 1985, p. 174). Reaction
formation is more 'energy-consuming' than sublimation.
Reich (1929) observed that, while reaction formation is
'cramped and compulsive', sublimation is 'free and flowing'
(p. 141). Reaction formation involves "a reversal of the
direction of the instinct", whereas, in sublimation, "the
instinct is simply taken over by the ego and diverted to a
different goal" (p. 141). When activity that represents a
reaction formation is interrupted, "restlessness will appear
sooner or later which may increase to irritability or even
anxiety", whereas sublimation-related work can be

interrupted for quite a while (p. 141). Moreover, "reactive achievements are less successful socially than sublimated ones" (Reich, 1929, p. 141). "Reaction-formation restricts and confines the manifestations of the personality", giving it "an appearance of spuriousness, hypocrisy, or camouflage" (Flugel, 1945, p. 70). This may undermine the purpose of reaction formation, which is to avoid others' criticism and ridicule and to thereby protect one's self-evaluation and self-esteem.

'Ego restriction' is another characterological defense mechanism described by Anna Freud. The ego refuses to re-enter a situation in which anxiety was felt, a situation that is henceforth seen as 'dangerous' (A. Freud, 1937, p. 93). The realization of inferiority would have, in primate evolution, become associated with vulnerability, dangerousness, and anxiety. Children avoid situations in which they are inferior. What is avoided is 'an unpleasant realization of inferiority' (Sandler, 1985, p. 356). Neurotic persons, in particular, avoid engaging in "tasks that might, so they fear, if they fail them, injure their self-esteem and interfere with their struggle for personal superiority, their struggle to be first" (Adler, 1938, p. 130). On the other hand, we may avoid a competitive activity despite our superiority in this activity because of a "fear of the aggression which might arise from the envy of competitors" (Sandler, 1985, p. 356). Restriction of the ego can be associated with an attitude of regarding ourselves "as dabblers, amateurs, dilettanti, or beginners" (p. 70), so as to not be subjected to others' ridicule or condemnation (or even to self-blame and self-condemnation) for our inferior performances (Flugel, 1945, p. 71). Ego restriction does not reduce anxiety but prevents it from arising. If situations that potentially give rise to anxiety cannot be avoided, if "protection against anxiety through restriction of the ego" is lost, then neurosis may be precipitated (Sandler, 1985, p. 358).

Compensation is a related defense mechanism; and it is also closely related to Adler's concepts of 'inferiority complex' and 'striving for superiority'. Compensation refers to the person's efforts to offset any deficiencies in his physical or intellectual endowment (Laughlin, 1970, p. 26). Compensation "may be the result of actual inferiorities, deficiencies, and losses, or it may follow purely subjective and even quite unrealistic feelings of this nature" (p. 25). Power or prestige and athletic skills or scholastic achievements compensate for perceived or actual personal deficiencies. Compensation, being an ego defense, operates outside conscious awareness; although it may be "achieved primarily through conscious efforts" (Laughlin, 1970, p. 24), in which case the notion of compensation more clearly overlaps with concepts of striving for superiority (Adler), self-actualization, or pursuit of the idealized self-image (Horney). Like other characterological defenses, compensation is motivated by the need for acceptance and love; it reflects a conscious or unconscious desires to attain approval, acceptance, or love from others and thereby to bolster one's self-esteem (Laughlin, 1970, p. 25).

2.4 Vanity and Self-Aggrandizement

Narcissistic personalities are extremely self-centered and show a high degree of self-references in interactions with other people (Kernberg, 1970). Narcissistic persons have both an inflated concept of themselves and an extraordinary need for approval from others (Kernberg, 1970). They are overly dependent on others' approval and overly focused on others' good opinion. Persons with a narcissistic personality disturbance, being unable to maintain their self-esteem at normal levels, have an "intense hunger for a powerful external supplier of self-esteem and other forms of emotional sustenance in the narcissistic realm" (Kohut, 1971, p. 17). Narcissistic persons tend to be "vain, boastful, and intemperately assertive with regard to their grandiose claims" (Kohut, 1971, p. 178). They may openly demand special prerogatives, their 'narcissistic expectation' being that the world owes them devotion and glory and that "devotion or glory can be obtained without effort and initiative of [their] own" (Horney, 1939, p. 95). In vanity, grandiose claims or expectations are combined with strong (poorly restrained) exhibitionistic impulses. Vanity makes people "think constantly either of themselves, or of other people's opinions of them" (Adler, 1927, p. 157). Vein persons greedily seek public acclaim; they always try to be at centre stage in their social circle, "to be constantly in the limelight" (p. 161). Vanity can also be seen "in those people who dress conspicuously, or with self-importance, who deck themselves out for a brave show, in the same way that primitive peoples made an exhibition of themselves by wearing an especially long feather in their hair when they have reached a certain degree of pride and honour" (p. 172). Vanity, an expression of exhibitionism, is associated with lack of shame (Adler, 1927), shame being the counterpart and regulator of exhibitionism.

Excessive reliance on others as sources of approval is coupled with weakness of libidinal object relations (Fenichel, 1946; Annie Reich, 1960). What is distinctly lacking in pathological narcissism is the capacity to genuinely care for, love, or become involved with others (Kernberg, 1970). The narcissistic person is involved with others only insofar as use can be made of them as sources of approval and admiration, or insofar as the qualities of those who are idealized and admired can be owned by the narcissistic person (Kernberg, 1970). Adler (1927), too, thought that 'social feeling' (i.e., the capacity to love and care, to make libidinal investments in others) is insufficiently developed in persons who are always "preoccupied with the question of what other people think about them and with the impression that they make on the world", people, that is, who pursue their goal of power and superiority "with great intensity and violence" and "live their lives in the expectation of great triumphs" (p. 157). Vanity, if excessive, cannot coexist with concern for others, although "every human being is vain to some degree" (p. 157) (and "it is impossible to divorce ourselves entirely from a certain degree of vanity" [p. 159]). In persons whose capacity to love and care is relatively intact, there is only a muted striving for personal power and prestige (p. 149), and the goal of superiority "grows in secret and hides behind an acceptable façade" (Adler, 1927, p. 138). Vein persons, on the other hand, are "occupied solely with the thought of what they must still achieve, still possess, in order to be happy" (p. 172). Vanity, if excessive, is a cause of permanent unhappiness; it "makes people constantly dissatisfied, and robs them of their rest and sleep" (Adler, 1927, p. 169). Narcissistic persons, as Kohut (1971) pointed out, experience a chronic sense of dullness and passivity, but they feel suddenly happy and alive (and recover "a sense of deep and lively participation in the world") when they have had the benefit of others' interest or praise (p. 17).

Weakness of libidinal object relations not only leads to excessive use of others as selfobjects (Kohut) or as 'extensions of oneself' (Kernberg, 1970) but also gives rise to undue concern with the self. There is overinvestment in the self as a structure that is used defensively to compensate for tenuous object relations. The self is more needed (and has to be hypercathected) when external object relations are fragmentary and not supported by libidinal investments, and when it is also less reliable as a mechanism compensating for absent external appraisals (when the self as an internal self-esteem-regulating structure is defective [Kohut, 1971]). To say, with Kohut and Wolf (1978), that the self of narcissistic persons is more prone to fragmentation is equivalent to stating that these persons feel inferior and vulnerable and less secure in themselves. Owing to their inner insecurity, they have a heightened need for external appraisals; and their weak self is incapable of covering over their insecurity and anxiety internally. Narcissistically chosen and used objects can stabilize a weak self-representation temporarily, but the self remains unstable if it is not also supported by a loving internal object, the kind superego; and it is this internal object that normally stabilizes the self and moderates its dependence on external objects. Instability of the self (lasting enfeeblement of the self) stems, according to

Kohut and Wolf (1978), from failure by early selfobjects (parents, i.e. precursors of the superego) to consistently and realistically gratify the child's mirroring and idealizing needs. Persistent failures of selfobjects in early development cause primary disturbances of the self, including narcissistic personality and behavior disorders (Kohut & Wolf, 1978).

Vanity and frank exhibitionism, undue craving for praise and relentless efforts to validate a grandiose self (including the striving for superiority), and self-inflation in fantasy are somewhat different but closely related manifestations of inferiority and insecurity (basic anxiety). Self-inflation, that is, the tendency to aggrandize oneself and appear unduly significant to oneself, cannot be easily separated from an undue craving for admiration from others (Horney, 1939, p. 90). Self-inflation or -aggrandizement allows the narcissistic person to escape "the painful feeling of nothingness by molding himself in fancy into something outstanding" (p. 92); it consoles him for not being loved and appreciated (p. 93). Self-inflation, which Horney (1939) considered to be central to the concept of narcissism, is one of the mechanisms by which the person attains safety and counters the vital danger implicit in 'losing caste' (p. 198). Self-inflation can be more or less concealed. Self-aggrandizement in fantasy has to be concealed inasmuch as any revelation to others of one's grandiosity would provoke intense shame and anxiety (Kohut, 1971). Self-aggrandizement tenuously prevents the collapse of the self, but, if shame is a strong possibility, would do so only for as long as grandiosity is not admitted to others or even consciously to oneself. Similarly, exhibitionistic impulses may be concealed (by way of reaction formation) or sublimated; and they usually are if the person lacks assertiveness (Kohut) or 'courage' (Adler).

Adler (1927) pointed out that "exhibitions of vanity are not considered good form" (p. 157); "people who demand special treatment are usually either antagonized or ridiculed" (p. 162). Vein persons, insofar as they realize how they are estranging themselves from society, have to "make every attempt to camouflage the overt signs of their vanity", which they may do by "dressing sloppily and neglecting their appearance in order to indicate that they are not vain" (p. 160). Adler (1927) saw that "there is a type of modesty that is essentially vanity in disguise" (p. 157). Horney (1945, p. 167), too, argued that arrogance can be hidden behind overmodesty and apologetic behavior, depending on the measure of aggression available in the personality. Hidden arrogance can reveal itself in the hurt a person feels when the prerogatives he demands are not spontaneously granted to him (Horney, 1945, p. 167). Some narcissistic persons cover their extreme need for others' approving and echoing responses by a display of independence and self-sufficiency (Kohut, 1971, p. 293). If the striving for superiority is coupled with a low threshold for shame and heightened fear of ridicule, the insecure person leads an isolated and eccentric life, whilst maintaining his fiction of greatness in fantasy (Adler, 1927, p. 162). In the schizoid personality, exhibitionistic impulses are profoundly inhibited and may not even find an indirect outlet (without the risk of further alienation from others); then self-aggrandizement in fantasy remains the only means of securing the self.

2.5 Affective Manifestations

Exhibitionism can be expressed in a variety of forms, some adaptive and some neurotic. Identification with and enactment of a social role allows for socially acceptable (sublimated) expressions of exhibitionistic impulses. Schizoid persons, who have difficulties expressing emotions in a social context and especially 'giving in the emotional sense' (p. 14), are nevertheless "able to express quite a lot of feeling and to make what appear to be quite impressive social contacts" by "playing a role or acting an adopted part" (Fairbairn, 1952, p. 16). Exhibitionism drives the tendency of schizoid persons to adopt or play roles. Pursuits of the artistic kind, too, provide a means of exhibitionistic expression without involving direct social contact (Fairbairn, 1952, p. 16). Exhibitionism is allied with enthusiasm, which in itself can be neurotic. Enthusiasm is neurotic "when enthusiastic reactions become a character trait, a habitual, chronic response", and when these reactions are "not as selective and the occasions which evoke enthusiasm are less worthy" (Greenson, 1962, p. 182). Neurotic enthusiasm serves to ward off a 'painful underlying state' (Greenson, 1962, p. 176). In an attempt to cover over a defect, acquired in childhood, in the psychological structure of the self, a person may "be overly enthusiastic, dramatic, and excessively intense" when responding to everyday events, "often to the embarrassment of those around" him (Kohut, 1977, p. 5). Behind such 'pseudovitality', there lies "low self-esteem and depression – a deep sense of uncared-for worthlessness and rejection, an incessant hunger for response, a yearning for reassurance" (p. 5). Similarly, in an attempt to cover over a primary defect in the self, resulting from empathy failures of early selfobjects, there may be "an intense devotion to romanticized cultural – esthetic, religious, political, etc. – aims" (romanticized ideals that feature in "an excited, overly enthusiastic, hyperidealistic adolescence devoid of meaningful interpersonal attachments" and that "do not recede into the background when the individual reaches adulthood") (Kohut, 1977, pp. 5-6).

Neurotic persons have a stringent need to impress others and to be admired and respected (Horney, 1937, p. 171). Neurotic persons, in order to gain prestige and maintain self-esteem, "have to be able to talk about the latest books and plays, and to know prominent people" (p. 172). Neurotic persons "have to know everything better than anyone else"; they "want to be right all the time, and are irritated at being proved wrong, even if only in an insignificant detail" (Horney, 1937, p. 168). Some persons have a tendency to assume "the role of the all-knowing one, appearing to understand everything immediately" (Adler, 1927, p. 205), thereby seeking to assure themselves of the approval of teachers or leaders (acting as superego projections). Those with a narcissistic personality disturbance may yield to the pressure of their grandiose self and resort to lying, whereby they ascribe some great achievement to the self (Kohut, 1971, p. 110). This is reminiscent of, and lies on a spectrum with, self-aggrandizement in fantasy. Alternatively, "the persistent demand of the grandiose self" forces narcissistic persons "to respond with unusual performance" (p. 112), hoping they would thus "live up to the assertions of the grandiose self concept on which they have become fixated" (Kohut, 1971, p. 112). This emphasizes the continuity with healthy ambition in pursuit of realistic goals (Kohut's 'bipolar self') but also brings us back to Adler's concept of striving for superiority and Horney's concept of realization of the idealized self. The neurotic person, creating an idealized image of himself and wanting "to appear, both to himself and others, different from what he really is", is afraid that "others should find him out", should find out that he is a 'bluff' (Horney, 1945, p. 148). His 'fear of exposure', ultimately a fear of disregard, humiliation, and ridicule, is provoked by any situation that might make him conspicuous (such as a test situation or social gathering) (p. 149). The fear of exposure (as a bluff) and of consequential humiliation and embarrassment can manifest as shyness, particularly in any new situation, or as "weariness in the face of being liked or appreciated" (p. 150). The fear of exposure is related to blushing or the fear of blushing (Horney, 1945, p. 149), further suggesting that presentation to others of an 'idealized image' is a form of exhibitionism.

Shame is felt when exhibitionistic efforts are thwarted, when attempts to enlist others' participation in one's exhibitionism are not met with approval (Kohut, 1966). The narcissistic person is flooded with shame when the acclaim he expects to receive in a particular situation is not forthcoming or when he recalls a situation in which he "told a joke which turned out to be out of place" or "talked too much about himself in company" (Kohut, 1971, p. 230). Shame curtails exhibitionism and engenders an impulse to conceal oneself (Fenichel, 1946; Kohut, 1971; Morrison,

1983). Feeling ashamed implies that "one is completely exposed and conscious of being looked at", while one is "not ready to be visible" (Erikson, 1950, p. 227). When feeling ashamed, one does not want to be seen, because being looked at then is "automatically equated with being despised" (Fenichel, 1946, p. 139). Shame reflects the deeper danger inherent in one's exposure to others (Morrison, 1983). Shame arises when efforts to actualize one's grandiose or idealized self (ego ideal) are thwarted, when the concealed belief in one's grandiosity collapses and one's inferiority becomes evident to oneself and others. Shame, as Morrison (1983) understood it, represents a tension between self and ego ideal. Shame can also arise from an unfavorable comparison between self and peers, whenever a perceived deficiency or inferiority is felt to potentially invite hostility from others. Guilt is a common defense against shame (Erikson, 1950, p. 227), although not in narcissistic persons per se who are not easily "swayed by guilt feelings (they are not inclined to react unduly to the pressure exerted by their idealized superego)" (Kohut, 1971, p. 232). When shame is absorbed by guilt, the self (ego) seeks to appease the superego by way of self-recrimination, in the hope of being accepted back into the loving protection of the superego. As Erikson (1950) put it, 'visual shame' is followed (and covered over) by 'auditory guilt', "which is a sense of badness to be had all by oneself when nobody watches and when everything is quiet – except the voice of the superego" (p. 227). Guilty feelings and their accompanying self-recriminations are acknowledged and expressed very readily, whereas the state of being found out, when one feels sincerely regretful or ashamed of something, is painful; and it is even more painful to express this feeling to someone else (Horney, 1937, p. 233).

Shame (the state in which one feels like one is despised and ostracized) is closely related to social anxiety, which is a constant *fear* of being criticized and ostracized (Fenichel, 1946, p. 518). Human anxiety, as Erikson (1950) stated, is a "fear of invasion by vast and vague forces", of "strangling encirclement" by enemies, and of "devastating loss of face before all-surrounding, mocking audiences" (pp. 365-366). Erythrophobia (fear of blushing), stuttering when speaking in public, stage fright, and other social fears are based upon inhibited exhibitionistic impulses (Fenichel, 1946, p. 316). Erythrophobia and stage fright represent an unconscious "idea that what is done to protect the person's self-esteem against danger may result in the opposite, in his complete annihilation" (p. 201). Socially anxious persons are unable to endure a state of not being loved or not being appreciated, because their self-esteem is weak (Fenichel, 1946, p. 519), which, in turn, is due to a lack of appreciation given to them by early love objects (Schilder, 1951, p. 94). Their self-esteem or self-appreciation depends, to a high degree, on continuous narcissistic supplies from outside; yet they intensely fear rejection and thus have to suppress aggressive and exhibitionistic strivings normally employed to elicit these supplies (Fenichel, 1946, p. 520; Schilder, 1951, p. 91). Hypochondriasis can be regarded as a state of 'heightened narcissistic-exhibitionistic tension', which arises when discharge of exhibitionistic impulses is incomplete (Kohut, 1966). Diffuse anxiety at times of separation from, or loss of control over, selfobjects (anxiety that accompanies fragmentation of the self) flows into hypochondriacal worries, preoccupations with various physical sensations and minor physical defects (Kohut, 1977, pp. 155-156). Hypochondriacal worries and symptoms of somatization can, at the same time, express frustrations in efforts to control selfobjects and pave the way for solicitation of narcissistic sustenance by alternative means, namely through illness behavior.

2.6 Summary

Compliance wards off intraspecific aggression and thereby defends against persecutory or paranoid anxiety; but compliance is also a *precondition* for the solicitation of narcissistic sustenance (approval and praise). Exhibitionistic behaviors *actively* solicit narcissistic supplies in the form of approval or praise. Exhibitionism is concerned not simply with attracting attention from others but with attracting positive (narcissistically nourishing) attention. Exhibitionistic means of attracting others' positive attention (and thereby upholding the narcissistic homeostasis) have to be modified, that is, they have to take into account (and *comply* with) external 'reality'. Characterological defenses, like sublimation, reaction formation, ego restriction, or compensation, are motivated by the need for others' approval. Sublimation, in particular, ensures that the safety-enhancing effect of being in the focus of others' attention is not offset by greater vulnerability to others' ridicule, a vulnerability that is signaled to the ego by feelings of paranoid anxiety and shame. Shame is an imminent expectation of being ridiculed and victimized by the group; the fear is that actions that were meant to elicit others' approval may actually trigger the group's concerted ridicule and hostility. Shame arises from a perceived failure of exhibitionistic efforts (to gratify one's need for others' approval) or from an unfavorable comparison between self and others (competitors for the attention of the leader). Shame curtails exhibitionism and urges the subject to hide from others (or blend in with others and prioritize compliance). Guilt (self-recrimination) helps to overcome shame (and avert the prospect of annihilation by the group) by ensuring oneself of the protection of the superego (or leader).

The ego ideal addresses the narcissistic vulnerability associated with unrestrained exhibitionism. The 'ego ideal', 'pole of ideals' of the bipolar self, 'ideal self', or 'idealized self' represents the goal toward which the personality as a whole strives; it safely guides the person's exhibitionistic efforts to recapture the security that is associated with being loved or accepted by, of being considered worthy of attention and approval from, the maternal object, the superego, or one of its external representatives, especially the leader of the group. The ego ideal incorporates the standards of those who have the greatest respect in the group; and it so aids the propagation of cultural norms and achievements. Through approximation of the ego ideal (brining to bear our skills, talents, and knowledge), we enhance our approvability in the eyes of the superego or leader and safeguard the goodwill of the superego or leader, apart from averting the hostility of the group.   Approximation of the ego ideal, through accumulation of possessions or prestige, allows for merger experiences with the superego, the internal representative of the primary maternal object. The striving for superiority expresses not only a wish for acceptance by the internalized maternal object but also an attempt to deny this very need for acceptance and thus to deny one's dependence on others and more particularly on the group and its leader. There are two principle approaches to harnessing the safety promised by the ego ideal or idealized self-image: self-actualization (through the more or less compulsive pursuit of reality-orientated ambitions) and self-deception, whereby the latter can take place in imagery alone (in a state of detachment) or consists in deceiving oneself about the nature and extent of others' interest in oneself (arrogance).

The notions the individual has about his self (ego) and his pride in his individuality compensate for a lack of firm rootedness in the group and/or an insecure relationship to the mother earlier in life. The less firmly the individual is embedded in the group and the less he had felt, earlier in life, accepted by the mother, the more his self will have to be bolstered by self-deception or the more he will strive for superiority (the more, that is, he will be driven to actualize a security-promising image of himself [i.e., the ego ideal, ideal self, or idealized image] against which he has to measure his actual self). Maintaining the semblance of one's ego ideal involves self-control, which can be excessive and compulsive, giving the personality a cast of artificiality ('false self' or persona). In order to gratify the need for approval, spontaneous exhibitionistic, self-assertive, and affectionate tendencies (expression of the 'true self') may have to be suppressed, at least in some situations, especially if a heightened need for approval is coupled with heightened fear of rejection, disapproval, or ridicule (as can be seen in neurotic inhibitions). Exhibitionistic tendencies would then seek indirect expression, even in somatic symptoms. *Restrictive* self-control, which is mostly negatively motivated (avoidance of negative attention), may collaborate (in one's efforts to approximate one's ego ideal) to varying degrees with ambition and arrogance (representing more *expansive* attitudes, variants of exhibitionism), giving rise to different personality types. Both, excessive self-control (self-definition) and arrogance (vanity), are associated with weakness of the self (equivalent to an inner sense of insecurity, to doubts about one's acceptability to the primary object or superego). Self-aggrandizement (whether in fantasy or in interaction with reality) prevents the collapse of a weak self, but self-aggrandizement may in itself have to be concealed from others and from oneself if the experience of shame is a strong possibility. Arrogance and vanity can be camouflaged (by way of reaction formation) behind excessive modesty, apologetic behavior, and self-neglect, thus preventing one's own and others' awareness of one's grandiosity. If restrictive self-control is the predominant approach to ensuring acceptance by the superego, then omnipotence and grandiosity (albeit present) are largely hidden from awareness.

Chapter 3

Assertiveness and Aggressive Control

The child's healthy exhibitionism is complemented by healthy assertiveness vis-à-vis mirroring selfobjects. Aggression, as Kohut (1977) saw it, "is, from the beginning, a constituent of the child's assertiveness, and under normal circumstances it remains alloyed to the assertiveness of the adult's mature self" (p. 116). Aggression is "a constituent of the firmness and security with which [the child or adult] makes his demands vis-à-vis self-objects who provide for him a milieu of (average) empathic responsiveness" (p. 118). Aggression, being an integral part of socially adaptive, nondestructive assertiveness, is aimed at controlling the emotional responsiveness of the selfobject, "whenever optimal frustrations (nontraumatic delays of the empathic responses of the self-object) are experienced" (Kohut, 1977, p. 121). Angry behavior, as Bowlby (1973) recognized, has the function of coercion and, when expressed toward a parent or partner, acts to promote, not disrupt, the bond (p. 248). Anger in the form of reproachful behavior toward the attachment object has the function of discouraging her from straying away again or occupying herself in other matters (Bowlby, 1973). Anger becomes dysfunctional, manifesting as hatred, "whenever a person, child or adult, becomes so intensely and/or persistently angry with his partner that the bond between them is weakened, instead of strengthened" (pp. 248-249), "whenever aggressive thoughts or acts cross the narrow boundary between being deterrent and being revengeful" (Bowlby, 1973, p. 249). In Kohut's (1977) terms, destructiveness emerges as a 'disintegration product' when 'empathy failures' of selfobjects cause the 'psychological configuration' of assertiveness to break down (pp. 114-115). Destructiveness is "the result of the failure of the self-object environment to meet the child's need for optimal ... empathic responses" (p. 116). Chronic narcissistic rage becomes embedded in the personality if the phase-appropriate need for omnipotent control over the selfobject was consistently or traumatically frustrated in childhood (Kohut, 1977, p. 121).

Aggression, according to the concept of 'neutralization' of instinctual energy, can be modified and diverted toward socially or culturally more acceptable or valued aims (Hartmann, 1964, p. 217). Neutralized aggression serves, for instance, the situationally appropriate solicitation of acceptance and approval, that is, of narcissistic sustenance for the self. Aggression, usually in its neutralized from, is used to maintain or restore one's dominance position in the group and others' *respect* for oneself and thereby to maintain one's safety (as reflected in self-esteem). The striving to dominate others is closely related to 'the urge to impress' others (Hass, 1968). The urge to dominate prompts us to aspire to positions of social eminence and esteem (p. 138). A pleasurable sense of power is associated with impressing other people "with the attainment of superior positions, titles, decorations, and marks of distinction" (p. 205). Man surrounds himself with possessions and symbols designed to "accentuate the impression he makes on others and intensify his pleasurable sense of power", whereby "the desire to acquire them increases his willingness to work" (Hass, 1968, p. 183). By impressing others, by following his "striving for success, esteem, and power, for social acceptance and standing, for recognition, superiority, and admiration" (p. 179), he inhibits their aggression and maintains their respect. Failure to impress others has "a particularly corrosive effect"; it releases "contempt and repudiation on the part of others" (Hass, 1968, p. 179). Such failure also undermines access to narcissistic resources (which sustain the self) and endangers the self (facing it with the possibility of annihilation). If the individual does not receive the level of support he expects from others, then his own hostile aggression is a defensive strategy to handle 'fear concerning survival' (Heard & Lake,

1986, p. 436). Hostile aggression has, in the first place, the aim of subordinating others and restoring their respect, thus reestablishing territorial claims over narcissistic resources and reaffirming the self. Hostile (offensive) aggression can be maladaptive and, when failing to achieve its primary objective, turn into rage. The neurotic person, when his secret expectation of a return for the favors or generosity he has shown is not fulfilled, finds himself thrown back onto his basic anxiety and may, at the same time, experience rage (Horney, 1937, p. 135). Hostility, incited when efforts to solicit approval or affection are frustrated and when "self-esteem has been wounded by humiliation", can manifest as spitefulness or vindictiveness and take "the form of a desire to humiliate others" (Horney, 1937, p. 178).

The desire to outperform and defeat others is counterbalanced by a fear of others' envious and retaliatory actions. This is a fear of retaliation for the *ruthless* pursuit of ambitions for power, prestige, and possessions (Horney, 1937, p. 207). The neurotic person, in particular, "automatically assumes that others will feel just as much hurt and vindictive after a defeat as he does himself"; and so he is anxious about hurting them (p. 196). He fears that, if he were to annoy others, there would be "a final break; he expects to be dropped altogether, to be definitely spurned or hated" (Horney, 1937, p. 252). The neurotic person fears the begrudging envy of others (p. 214); he feels that "once he has shown an interest in success he is surrounded by a horde of persecuting enemies, who lie in wait to crush him at every sign of weakness or failure" (p. 211). He feels that, once he has made a mistake or shown some weakness, he will be the object of disrespect or ridicule (p. 224). Thus, basic anxiety (reflecting feelings of insignificance, helplessness, and insecurity) impels the person "to strive for and attain more and more strength and power in order to be safe" (p. 268). At the same time, the fear of others' envy and retaliation curtails spontaneous self-assertion, causes difficulties in criticizing others (or making accusations), and inhibits competitiveness

(Horney, 1937, p. 250). The neurotic person is afraid of people; and, insofar as he must present a rational front, he is prevented from feeling or venting any grievance against them (Horney, 1939, p. 242). Reproaches against others accumulate and are rechanneled onto the self, turning into self-reproaches (Horney, 1939, p. 242).

3.1 Loss and Ambivalence

Anger directed at a temporarily unresponsive attachment object aims to increase her responsiveness to one's attempts to induce her to display affectionate or attentive behaviors (Bowlby, 1973). When separation is permanent, that is, when object 'loss' has occurred (bereavement), pain and anger are similarly felt, but then "anger and aggressive behaviour are necessarily without function" (p. 247). During early phases of grieving, "a bereaved person usually does not believe that the loss can really be permanent" and continues to act as though it were still possible to find the lost person (p. 247); "the lost person is not infrequently held to be at least in part responsible for what has happened, in fact to have deserted", so that "anger comes to be directed against the lost person, as well as, of course, against any others thought to have played a part in the loss or in some way to be obstructing reunion" (Bowlby, 1973, pp. 247-248). Joffe and Sandler (1965) thought that "what is lost in object loss is ultimately a state of the self for which the object is a vehicle" (p. 178). When a love object is lost (and mental pain arises), "the affective value cathexis of the object is greatly increased, and attention is focused almost exclusively on the object because it is the key to the reattainment of the lost state of the self" (Joffe & Sandler, 1965, p. 159). The self, in order to avoid annihilation or disintegration anxiety, has to reattach itself to another representative of the primary object, either in reality or in fantasy. The self is inseparably bound to the primary object or any later representative of the primary object; and, if the self cannot reattach itself to a conscious derivative of the primary object, it seeks consolation in the illusion of omnipotence, wherein an omnipotent self is bound up with the unconscious omnipotent object (an aspect of the superego).

The pain and anger experienced after loss of an external object is related to narcissistic injury and rage felt in consequence of a slight or rejection. Tenuous links to objects that only serve as selfobjects are easily broken. A slight or rebuff causes the grandiose self to collapse; and what is lost then temporarily is the link of the self to the omnipotent object. Hence, the loss also concerns an internal object, unconsciously. Narcissistic injury, when acceptance or approval has been solicited but is not forthcoming, uncovers a "discrepancy between an ego created in fantasy and the actual person" (Federn, 1952, p. 313), "a discrepancy between the actual state of the self on the one hand and an ideal state of well-being on the other" (Joffe & Sandler, 1965, p. 156). The pain and anger arising from a wound to self-esteem may be rationalized as disappointment. Narcissistic persons are hypersensitive to slights; they tend to react with rage and revengeful fantasies when their need for constant admiration is frustrated, when their assertive and exhibitionistic outreach for narcissistic sustenance runs aground and control over selfobjects is lost (Kohut, 1977, pp. 259-262). Aggression is primarily aimed at restoring the empathic responsiveness of the selfobject surround, but narcissistic rage will be counterproductive. In those with a 'narcissistic behavior disorder' (pp. 193-195), perverse or delinquent behaviors, unconsciously designed to force mirroring responses from the selfobject milieu, "are only *one* step removed from the underlying defect in self-esteem" (Kohut, 1977, p. 195).

Mourning the loss of an object is not dissimilar to the state depression, in which there is a loss of control over selfobjects. Mania is a defensive mode that aims to prevent depressive feelings of loss and guilt (Klein, 1940). If the developmentally normal 'depressive position' was not overcome successfully in early childhood, then, in later life, loss of (control over) any external object reawakens 'depressive anxieties' about losing the 'good internal object' (counterpart of a secure self). It is the loss of the good internal object that is unconsciously mourned in mania, whereby excessive anxiety about the possibility of losing the good *internal* object (leading to insecurity of the self) is associated with an inability to form *real* object relations (involving investment of libido and the sharing of 'social feeling'). In the reawakened depressive position, the person gains insight into his lack of real object relations (believing unconsciously that he has 'destroyed' his objects) and hence into his fundamental loneliness (Riviere, 1936). Depressive anxieties, acting as a reminder of the 'damage' done to the good internal object (the introjected maternal object), can be avoided in part by *denying* dependency on external objects (Klein, 1940). Denial is a manic defense. Contempt for and triumph over external objects are further manic defenses (whereby these external objects are treated as mere selfobjects). Mania involves omnipotent control over selfobjects (objects that are all 'within' the person himself, in the sense that they only exist for himself [Riviere, 1936]), which aims to compensate for the lacking good internal object. The narcissistic person, too, unconsciously fears that any lessening of control over his external objects will reawaken depressive anxieties (Riviere, 1936). The narcissistic person is always close to being aware that he lacks real object relations; and his omnipotent defenses are an attempt to avoid the despair and depression that this insight would bring (Riviere, 1936).

The threat of losing a (narcissistically) needed external object increases efforts to control the object, whereby aggressive control over the object (ambivalent attack) not only fosters an omnipotent sense of independence but also, in a vicious circle, deepens the feeling of guilt, which, in turn, is defended against by denial of dependency on and sadistic triumph over the object (Segal, 1973). Underneath the person's attitude of omnipotent control, there is "a craving for absolute bliss in complete union with a perfect object for ever and ever", which is "bound up with an uncontrollable and insupportable fury of disappointment" (Riviere, 1936). Persecutory fears, an aspect of the 'paranoid-schizoid position' into which the narcissistic or manic person readily regresses, strengthen manic mechanisms of defense. 'Bad' and persecutory objects need to be monitored constantly, so that the inner bad object (persecutory superego) can be manically subordinated. Attacks on external objects increase their destruction, "thereby deepening depressive anxieties and making the underlying depressive situation increasingly hopeless and persecutory" (Segal, 1973).

3.2 Jealousy and Envy

Jealousy illustrates the use of aggression for the purpose of binding an object in a relationship. Jealous persons "bind their partner with chains of love"; they "build a wall around their loved one" (Adler, 1927, p. 180). They feel they have an exclusive claim over their object's attention and love. Persons who are inclined to feeling jealous "are not able to love but need the feeling of being loved" (Fenichel, 1946, p. 391). They have an intense fear of loss of love and long for an alternative, more secure source of narcissistic sustenance. Their longing for another object produces jealousy "on a projective basis; their longing for another partner is projected", so that they believe that it is their partner, not they themselves, who is looking for a new object (p. 391). The projection is reinforced by the special sensitivity of jealous persons to signs of their object's unfaithfulness (Fenichel, 1946).

While jealousy involves hostility against the attachment object, envy involves hostility toward a perceived rival (the third person in a triangular constellation) with whom the envious person competes for attention and love of the primary attachment object or of one of its external derivatives or, unconsciously, of the superego (internal omnipotent object). In either envy or jealousy, aggressive impulses are directed against obstacles perceived to be in the way of the striving for an exclusive object relation (in reality or unconscious fantasy). Envy is the feeling of resentment or hostility toward the good qualities or abilities of another person. An attempt or desire to spoil these qualities or abilities is fundamental to envy (Joseph, 1986). Persons who are preoccupied with the striving for power and domination and who "spend their time in measuring the success of others" (p. 181) (and comparing others' successes with their own achievements) become envious when they cannot gain superiority over others (Adler, 1927). They would then be "interested solely in taking things away from other people, in depriving them and putting them down" (Adler, 1927, p. 182). As Horney (1950) argued, neurotic persons have 'grandiose claims', which are responsible for their chronic smoldering envy and discontent (p. 47). Their self-esteem depends on their power, possessions, and prestige; and they enter a state of misery when they fail "to have the one advantage in which another person surpasses [them]" (as they compare themselves with the other person) (Horney, 1937, p. 183). While most of us "will feel some envy if others have certain advantages we should like to have ourselves" (p. 182), neurotic persons feel humiliated if they have to "give someone credit for something" (p. 196). In association with their attitude of begrudging envy, they have "a tendency to deprive others" (p. 180) and "to defeat or frustrate the efforts of others" (Horney, 1937, p. 193).

There are various defenses that the envious person can employ to protect himself against inacceptable and disapproved (and therefore egodystonic) feelings of envy. The envious person may devalue himself in order to increase the gap between himself and the envied person (a defense related to masochism) (Joseph, 1986). Alternatively, the envious person may regard the envied person as inferior and devalue the latter's qualities. The neurotic person readily devalues others' achievements and thereby "succeeds in assuaging his envy and discharging his resentment" (Horney, 1945, p. 203).

Another envious person may idealize the envied person, placing the latter on a pedestal and out of reach, just as the leader of a group would be (Joseph, 1986; Spillius, 1993). In a related defense, the envious person may identify himself with the envied person, allowing him to introject some of the latter's enviable qualities and vicariously possess these qualities. The envious person may project his envy (so that others appear envious) or stir up feelings of envy in others (by making them aware of his own outstanding qualities) (Joseph, 1986; Spillius, 1993). The envious person may restrict contact with others and avoid situations that can stimulate envy in himself. Competitiveness is a more adaptive defense against feelings of envy; in acquiring superior qualities or capacities (power, prestige, or possessions), the person avoids feeling envious toward others (Joseph, 1986).

The striving for prestige, wealth, and power is competitive; it can be all-consuming and openly aggressive. Neurotic persons not only have a tendency to deprive others; they are fearful that they themselves "will be cheated or exploited by others", that "someone will take advantage of [them], that money or ideas will be stolen from [them]" (Horney, 1937, p. 185). Whenever they do feel cheated or exploited, a disproportionate amount of anger is discharged (Horney, 1937, p. 186). Prestige, wealth, and power demarcate our narcissistic resources. In striving to obtain prestige, wealth, and power, we seek *unconsciously* to control the attention and love we continue to need from the primary maternal object (which was introjected in childhood and became the loving aspect of the superego). In this pursuit, we find ourselves in competition with others, with those who require narcissistic nourishment from the same central object representing their primary object. Competitors have to be outperformed or, if we fail to do so, their achievements have to be spoilt (enviously). Successful competitors (for the attention of a concrete or abstract central object, such as a leader or an organization) are envied; and, if such envy is egodystonic and arouses shame, defenses against envy have to be erected. Narcissistic persons may tolerate feelings of envy more easily and act on them unreservedly. Such envy was termed 'impenitent' (Spillius, 1993). Narcissistic persons need to be assured of their object's love more urgently; and their impenitent envy reflects, unconsciously, their mourning for the 'ideal parents' they wish they had had (and the love and attention they wish they had received) and also for the 'ideal self' they would like to have been (Spillius, 1993).

Persons who are easily consumed by envy are those "who are not concerned about making themselves useful to others" (Adler, 1927, p. 183). They are narcissists, in that they lack 'social feeling'. They can also be considered as antisocial personalities to the extent that they are not moved at all by the "fact that their actions cause suffering to others" (and they may even take pleasure in others' pain) (Adler, 1927, p. 183). Excessive envy prevents the person from having warm and trusting relationships. As a result, the envious person is likely to remain insecure, causing, in a vicious circle, increased disposition toward envy (Joseph, 1986). Others' happiness and their ability to care and love can become the subject of envious thoughts and actions. Others' happiness and "their "naïve" expectations of pleasure and joy" irritate the neurotic person (Horney, 1945, pp. 201-202). Begrudging envy together with an impulse to "trample on the joy of others" (p. 202), "to frustrate and to crush the spirit of others" (p. 202) arise in him when he sees others "love, create, enjoy, feel healthy and at ease, belong somewhere" (Horney, 1945, p. 201).

3.3 Righteousness and Sense of Entitlement

The striving for power, prestige, or possessions provides a channel through which 'repressed hostility' can be discharged (Horney, 1937, p. 166). In the striving for power, prestige, or possessions, whether neurotic or not, "a certain amount of hostility may be discharged in a non-destructive way" (pp. 174-175). Even when the need for power is compelling, it can be disguised in 'socially valuable or humanistic forms' and "does not necessarily appear openly as hostility toward others" (p. 174). However, when hostility is concealed in activities such as giving advice or taking the initiative, "the other persons ... will feel it and react either with submissiveness or with opposition" (p. 174). Hostility that was hitherto concealed and pressed into civilized forms can break out more openly when the neurotic person does not succeed in having his own way (p. 174). He would then show a "plain anger reaction to a lack of compliance" with his wishes and expectations (p. 169) or react angrily to others' failure to follow his advice (Horney, 1937). Adler (1927) observed that persons with deep inner insecurity (inferiority) (correlated with lack 'social feeling') not only engage in compensatory striving for superiority but also have a tendency to judge, criticize, and ridicule others (p. 162). Their sharp and critical manner, an expression of 'social hostility', allows them "to gain a feeling of superiority by degrading other people" ('depreciation complex') (p. 163). Their tendency to degrade others is associated with their insistence on always being right. They go to great lengths to prove themselves right and to prove others to be in the wrong (Adler, 1927, p. 198). Horney (1945) suggested that "a combination of predominant aggressive trends and detachment is the most fertile soil for the development of rigid rightness; and the nearer to the surface the aggression, the more militant the rightness" (p. 138). The 'aggressive type' of neurotic personality "seems to have an unusual capacity for definite opinions"; "his opinions will often have a dogmatic or even fanatic character" (Horney, 1945, p. 170).

Bursten (1973) argued that argumentativeness, critical suspiciousness, and jealousy, shown by persons with a 'paranoid type' of narcissistic personality structure, reflect a sense of disappointment or betrayal when reunion with an omnipotent object cannot be achieved. Argumentativeness, critical suspiciousness, and jealousy (comprising a 'mode of narcissistic repair') betray the person's need to be the 'special selected one' in the eyes of the omnipotent object (Bursten, 1973). A deep sense of inadequacy (which renders the person unacceptable in the eyes of the omnipotent object) and related shame are counteracted by externalizations and projections. The paranoid narcissist constantly looks for shameful conduct in others, so as to affirm and support his projections. If others can be seen as inadequate and shameful, the paranoid narcissist feels more acceptable again to the omnipotent object with which he unconsciously seeks to reunite (Bursten, 1973).

Being driven and *compelled* to actualize his idealized self-image (turn himself into the glorified image of himself), the neurotic person is "liable to expect an unreasonable amount from others"; he makes unreasonable claims on others, to the fulfillment of which he feels entitled ('grandiose claims') (Horney, 1950, p. 370). The neurotic person feels "entitled to be treated by others, or by fate, in accord with his grandiose notions about himself", "entitled to special attention, consideration, deference on part of others" (p. 41), "entitled never to be criticized, doubted, or questioned" (p. 43), "entitled to everything that is important to him" (p. 42). He insists on having special rights and takes benefits "accruing from laws or regulations ... for granted" (p. 44). Not only does he assert exceptional rights for himself, he may also adopt "a right, a title, which in reality does not exist" (Horney, 1950, p. 42). The 'sense of entitlement' is a sense of having special rights and of being entitled to break the law (Moses, 1989). The person who has a sense of entitlement "holds tenaciously to the conviction that his behavior is correct, appropriate, and adequate" (p. 489). Any challenge to his entitlement, to "his view of himself as someone with rights of special entitlement", can lead to self-righteous indignation (p. 492) and 'the righteous rage of entitlement' (Moses, 1989, p. 488). In other words, if the neurotic person's grandiose claims are not satisfied and his exceptional rights not respected, if others do not cater to his grandiose illusions, then he experiences a deep sense of unfairness, and he can become furiously indignant (Horney, 1950, pp. 41-42). Nonfulfilment of grandiose claims "is felt as an unfair frustration, as an offense about which we have a right to feel indignant" (p. 42). Hitherto inhibited and neutralized aggression can thus be released directly. Severe reactions to frustration "are indicated by the terror of doom and disgrace" and by rage at self and others (Horney, 1950, p. 31).

The sense of entitlement derives from the child's demands for the total and undivided attention of the mother (Moses, 1989, p. 485). Children gradually give up their sense of entitlement; but it persists in ameliorated form into adulthood as the making of "demands which are appropriately one's right" (p. 486). The sense of entitlement is excessive in persons who believe that they were subjected in their childhood to unjust deprivation and suffering. The sense of entitlement, when it is excessive, is a feature of the narcissistic personality and coexists with shamelessness. Shame and the sense of entitlement "are in some way opposite sides of the same coin" (p. 484), both being "closely related to the self, to narcissistic proclivities" (Moses, 1989, p. 483). Shame seems to curtail any sense of entitlement, much as it curtails exhibitionism. Inhibition or repression of the sense of entitlement, or of exhibitionism, can be maintained by reaction formation in the form of modesty. Indeed, the sense of entitlement may be unconscious, being hidden "behind a cloak of marked modesty" (Moses, 1989, p. 486).

3.4 Vindictiveness and Manipulativeness

Having been wronged, the neurotic person may become vindictive. Experiencing an impulse to get back at others and the desire to vindictively triumph over them, he "tries through hardhitting accusations to enforce their compliance" and "may ruin [them] with insatiable claims" (Horney, 1950, p. 55). The neurotic person, having been wronged by somebody, starts to ponder the hateful qualities of that person, so that that person "suddenly becomes untrustworthy, nasty, cruel, contemptible" (p. 56). Whenever the expression of overt anger and hostility is socially unacceptable and inhibited, "one will have to exaggerate the wrong done; one will then inadvertently build up a case against the offender that looks logic tight" (p. 56). Overemphasis on 'justice' having to be restored can be a camouflage for vindictiveness (p. 55). If anger can still not be expressed, the wronged neurotic person becomes despondent or plunges into misery and self-pity; or his anger may appear in psychosomatic symptoms (and his suffering would then become the medium to express reproaches) (Horney, 1950, pp. 56-57). The sadistic person is more than just vindictive. The sadistic person is habitually inclined "to tell others how stupid, worthless and contemptible they are and to make them feel like dust" (Horney, 1939, p. 220). By degrading others and striking at them "with righteous indignation from the height of [his] own infallibility" (Horney, 1939, p. 220), he "gives himself a feeling of superiority" (Horney, 1945, p. 206). He "gains a stimulating feeling of power over them" (p. 206); and his feeling of strength and pride reinforces his unconscious feeling of omnipotence (p. 207). His triumphant "elation at being able to do with others as he pleases" thus "lessens his own sense of barrenness" and "obscures his own hopeless defeat" (p. 207). Sadism, as a neurotic trend, is a defense against deep-seated inferiority and self-contempt (Horney, 1945).

On account of feelings of inferiority and insecurity, aggressive criminals are more likely to perceive social threats, such as in the form of another's real or imaginary minor aggression, and react to these with counteraggression (Schilder, 1951, p. 215). Criminal aggression is not only "a reaction to an immediate situation" but also "to a situation in childhood" (p. 215), a childhood that has imbued the individual with feelings of inferiority and insecurity. Schilder (1951) proposed that "aggressive action takes place when the individual feels restricted in his power to achieve an adequate mastery of the situation" (p. 219). Acts of aggression help the individual to restore his 'threatened masculinity', that is, to reassert control over his fate and environment; "the assault becomes a symbol for masculinity regained" (p. 215). The regaining of masculinity also means that, through the aggressive act, the aggressive criminal restores his *prestige* (p. 218) and thus his self-esteem (narcissistic homeostasis). The individual expects that the social environment, once mastery over it has been reestablished, will provide him with the needed approval and respect (narcissistic sustenance). As Schilder (1951) put it, an "aggressive impulse has a much greater chance of becoming criminal action when the criminal action can reckon with open or tacit approval of those social factors which play a part in the ego formation" (p. 217).

Vain (narcissistic) persons "who would like to rule others must first catch them in order to bind them to themselves" (Adler, 1927, p. 173). They may show an attitude of amiability, friendliness, or approachability in order to lull others into a sense of security and use them (narcissistically) so as to maintain their personal superiority (p. 173). They would however proceed to aggressively control those bound to themselves, thus removing their veil of amiability (Adler, 1927). In Bursten's (1973) classification, the 'manipulative type' of the narcissistic personality is characterized by manipulativeness and propensities for deception, superficial relationships, and contempt for others. Good manipulators can 'size up a situation' in order to influence others, to move them around in such a way that their own narcissistic needs will be gratified (Bursten, 1973). Manipulative personalities experience little guilt. Their 'mode of narcissistic repair', allowing them to overcome inner feelings of worthlessness, involves deception (lying) and aggressive competition aimed at defeating others. By defeating others, manipulative personalities prove their superiority and thereby their acceptability to the omnipotent object with which they seek to reunite (Bursten, 1973). 'Phallic narcissistic personalities', by contrast, achieve narcissistic repair (defense against shame associated with being weak or insignificant) by means of arrogance, aggressive competitiveness, and pseudomasculinity. Their need to be admired similarly reflects their wish for reunion with an omnipotent object. Phallic narcissists are prone to take risks, expecting they will be saved miraculously, and engage in acts of bravery for the sake of self-glorification. While their reliance on their omnipotent object is hidden in internalized structures, their grandiose self (the counterpart of the omnipotent object) is manifest in their ambition and competitiveness (Bursten, 1973). Phallic narcissists and manipulative narcissists have a relatively firm sense of self, reflecting their greater degree of individuation. They have more successfully internalized their sources of approval, whereas 'craving narcissists' depend on an almost continuous presence of their object (Bursten, 1973).

3.5 Passive Aggressiveness

Defiance, according to Rothstein (1979), represents a wish to experience victory over a frustrating parent, over an object that does not meet the child's narcissistic needs. It is an attempt by the child to force the object to be more loving and available, to be like the gratifying parent he remembers from earlier years; it the child's an attempt to recapture his original omnipotence, his lost sense of control over the maternal smile, that is, over narcissistic supplies (Rothstein, 1979). At the same time, defiance allows the child to indulge in omnipotent narcissistic self-preoccupation, thereby partially restoring his narcissistic balance in fantasy. Defiance is related to stubbornness. Stubbornness, too, is a passive type of aggressiveness (Fenichel, 1946, p. 279). Stubborn persons provoke others to be unjust, so that they can see themselves as being treated unfairly and, hence, attain "a feeling of moral superiority which is needed to increase their self-esteem" (p. 279). Stubborn persons are forever engaged in a "struggle for the maintenance or the restoration of self-esteem"; they are "filled with narcissistic needs, whose gratification is required to contradict some anxiety or guilt feeling" (Fenichel, 1946, pp. 279-280).

Hostility can be expressed indirectly through an 'appeal to justice' (Horney, 1937, p. 144). The neurotic person may use a traumatic experience or injury he suffered as a basis for demands for sympathetic treatment; he "may arouse feelings of guilt or obligation in order that his own demands may seem just" (p. 144). He may use his injury or illness implicitly as an accusation, as "a kind of living reproach, intended to arouse guilt feeling" in others and to make them willing to devote all their attention to him (p. 144). Neurotic persons may be willing to pay the price of suffering, "because in that way they are able to express accusations and demands without being aware of doing so, and hence are able to retain their feeling of righteousness" (Horney, 1937, pp. 145-146).

3.6 Reparation

Insufficient receipt of appreciation, approval, or praise from others confirms the person's suspicions about being inferior and unworthy of others' attention and regard. The person, especially the insecure person, must maintain his external sources of narcissistic sustenance and must control them assertively or even aggressively. Frustration of need stirs up aggressive impulses against the frustrating object, which is then regarded as 'bad'. Aggressive impulses against the object on whose love the person depends (a constellation known as 'ambivalence') have the potential to 'destroy' the object; they can result in the loss of the object. As Klein (1937) observed, the child (or adult) is afraid of losing the object on whom he depends; and he anticipates damage to this loved object as soon as anger wells up (p. 117). Fantasized destruction (more or less consciously) of the loved object causes 'unconscious guilt'. In order to preserve feelings of security, the child has to develop his capacity to keep loved objects safe and undamaged (Klein, 1937, p. 98). By way of showing love and care, damaged objects can be 'repaired', which ensures that these needed objects are not lost and that safety is retained; yet the ability to love (and show 'social interest') tends to be impaired precisely in those who are most insecure and most dependent on their objects' love and attention. Reparation, the making good of fantasized damage the individual has done to the object, assuages *unconscious guilt*. Hostility actually discharged (as opposed to fantasized) against the object gives rise to a more conscious feeling of guilt, a feeling that engenders the 'need for punishment' and prompts repentance (Flugel, 1945).

The need to make reparation may be allied with the need to be punished, in that the process of reparation may be laborious and arduous and may itself involve suffering (Flugel, 1945). Klein emphasized the role of useful work in assuaging unconscious feelings of guilt. Reparation involves not only the showing of love or care toward the object but also work aimed at regaining the object's approval. The work we carry out often aims at gaining the approval of the superego. Constructive work earns us the approval of colleagues and, importantly, recognition from authority figures onto whom the internal representative of the primary object is projected. As our dependence on the primary maternal object is carried over to a dependence on later objects, the need to make reparation to objects of our early life is "unconsciously carried over to the new objects of love and interest" (Klein, 1937, p. 117). We overcome unconscious feelings of guilt connected with aggressive fantasies by being considerate and helpful to those who stand for earlier objects which we had harmed in our unconscious fantasy (Klein, 1937).

3.7 Summary

Aggression, when neutralized and adapted to external reality, can help to strengthen the relationship with another person or keep the selfobject milieu, the source of narcissistic nourishment, under effective control. Aggression can be maladaptive and disintegrate into rage when the personal bond or the responsiveness of the selfobject surround cannot be restored. Loss of an external representative of the primary object is equivalent to loss of control over the selfobject surround, which, in turn, is equivalent to a deprivation in the state of self. If the loss is permanent, or if efforts to reestablish control over the selfobject surround are permanently frustrated, depression can arise. It can be deducted that a tendency to choose objects on a predominantly narcissistic basis (an inability to make libidinal investments in objects), leads not only to weak self-esteem but also to greater vulnerability to depression (as narcissistic resources would be both more needed and more fragile).

Aggression is woven into the competitive pursuit of possessions or prestige and into the betterment of one's social standing or ranking, all of which are closely related to one's control of narcissistic resources. Aggression is concealed within the cultural context, unless others' encroachment onto one's narcissistic resources and disregard of the 'rights' by which these resources are defined brings one's anger more clearly to the surface. Overt offensive aggression evolutionarily has the purpose of inducing submission in a challenger, whereby submissive behavior (respect) appeases and thus also narcissistically nourishes the aggressor. Aggression, in its overt variety, attains narcissistic sustenance precisely by suppressing the other's aggressive potential and thereby counteracting one's paranoid anxiety (the counterpart of narcissistic equilibrium). Neutralized aggression, which is integral to competitive pursuits and to the maintenance of one's social position, has the same effect. Through mutual induction of compliance and the display, toward each other, of compliance, individuals embedded in the hierarchical system of a group or society maintain each other's narcissistic balance (and keep each other's paranoid anxiety at bay), although occasionally perturbations will occur in such a dynamical system. The degree to which individuals' security depends on this interplay varies, depending on their endowment with social feeling or libidinal interest and on the level of security they experienced in their earliest relationships. Excessive insecurity on part of some individuals will translate into occasional challenges to the established order and contribute to perturbations in the system.

Overt aggression as a means of narcissistic repair may be justified in a particular cultural context; or the personality may have developed ways of making such aggression seem justified. The feeling of injustice or a strong longing for justice can be harnessed in association with more overtly aggressive behaviors that are geared toward subordinating others (inducing their submission) and restoring others' respect for oneself (and thereby ultimately restoring one's safety and protecting oneself against paranoid fears). If the cultural context prohibits overt expression of aggression, or if the fear of inciting, through one's aggression, others' retaliation is strong (if, in other words, the security-procuring effects of offensive aggression are likely offset by greater vulnerability to be attacked in return and potentially annihilated by the group), then aggression can be expressed in indirect ways. Defiance and stubbornness serve to provoke others to engage in unjust actions, which will, firstly, provide oneself with a sense of moral superiority, which, in turn, enhances one's sense of safety (through enhancing one's acceptability to the superego or leader), and which may also, secondly, provide oneself with a justifiable outlet for one's own aggression. One may use one's illness or an injury for the purpose of accusing others and arousing in them feelings of guilt, which would then prompt them to commit acts of repentance that can restore one's own righteousness and narcissistic equilibrium. Aggressive means of narcissistic repair are contrasted with compliant and submissive behaviors, pursing the same objective (but from the opposite end of the equation), and also with reparative behavior and work, aiming to regain the respect of others and of the superego, whereby work, too, would involve aggression in neutralized form (and may express, at the same time, the need for punishment, i.e., to be punished). Aggressive subordination of others may alternate in cycles, depending on mood and external cultural factors, with submissive and reparative efforts to maintain oneself in a relatively stable and thus *safe* position within the social hierarchy.

Chapter 4

Display of Helplessness
and Appeal to Pity

Infantile behaviors elicit caring responses from the mother. Care-seeking patterns also have the effect of inhibiting aggression (Moynihan, 1998). The infant's display of helplessness inhibits maternal aggression, apart from soliciting maternal care. Infantile behaviors, such as the display of helplessness, overlap with appeasement gestures. Appeasement gestures, including the display of vulnerability or weakness, similarly inhibit aggression in an opponent. Culturally or phylogenetically ritualized expressions of helplessness release, when displayed by an adult, the impulse to cherish and care in a partner or another group member. Humans who wish to elicit affectionate responses "relapse quite involuntarily into the role of a small child" (Eibl-Eibesfeldt, 1970, p. 148). Adult animals display infantile behavior patterns also when they seek to appease a conspecific (p. 113). Humans, for instance, display helplessness (and weakness) when the objective is to arouse pity in an opponent and so to curtail the opponent's aggression (Eibl-Eibesfeldt, 1970). Conversely, appeasement gestures, not all of which are related to infantilisms, facilitate the exchange of care-seeking and care-giving signals. In relationships, familiarity of partners with each other, too, inhibits aggression between them and thereby facilitates the exchange of care-seeking and care-giving gestures (Moynihan, 1998).

Narcissistic supplies in the form of others' caring and affectionate signals help to maintain self-esteem. Failure to elicit care-giving responses from others is injurious to self-esteem. If we fail to induce care-giving or affectionate responses in others, then pain or anxiety follows, which, in turn, predisposes us to display more overt 'attention-seeking' behavior or to engage in other 'self-defensive' behaviors (McCluskey, 2002). High anxiety and imminent 'self-fragmentation' bring about desperate attempts to provoke our social surround into showing attention and care toward us; but such desperate solicitations can precipitate antagonistic reverberations, which would worsen our anxiety and ultimately may push us into a state of (self-defensive) rage (Kohut, 1977; Wolf, 1988). Social recognition and respect regulate narcissistic homeostasis or self-esteem on a more abstract level. If we are to remain free of basic anxiety, then we have to manipulate our social environment into supplying narcissistic nourishment, and we have to do so in accordance with contingencies inherent in this environment. Thus, we tend to control our selfobject surround in situationally appropriate and culturally evolved ways. If self-esteem or self-cohesion were to be highly dependent on the explicit receipt of praise, approval, and attention, then we would have to use more concrete or primitive methods of obtaining narcissistic nourishment. 'Mirror-hungry personalities', described by Kohut and Wolf (1978), feel the compelling need to display themselves (in primitive exhibitionistic ways) in order to attract attention. They are "impelled to display themselves to evoke the attention of others, who through their admiring responses will perhaps counteract the experience of worthlessness" (Wolf, 1988, p. 78). Primitive methods of self-esteem regulation and mature or abstract methods, too, are employed more or less compulsively, depending on the extent to which the personality is endowed with 'social feeling' (capacity to genuinely show care for and love others). Persons with deep-seated feelings of inferiority (reflecting a lack of 'social feeling' or an enduring lack of genuineness and spontaneity) may be able to use mature methods of self-esteem regulation, but they would have to do so compulsively. Feelings of inferiority "may be covered up ... by a compulsive propensity ... to impress others and one's self with all sorts of attributes that lend prestige in our culture" (Horney, 1937, p. 37).

4.1 Persistence of Infantile Dependency

Klein (1937) described how features of our earliest relationships pervade our adult relationships. In our adult relationships, "the early wish to have one's mother or father all to oneself is still unconsciously active" (p. 75). When entering into a relationship with someone who has a parental attitude, we hope to gratify our earliest wishes for care and for reassurance and security (p. 76). The trust, protection, and help we, in turn, offer to our partner are based on *identification* with the trusting, protecting, and helpful parent from our own childhood. Both partners may alternately or simultaneously take the place of a parent or put the other into the role of a parent. Under favorable conditions, the relationship for both partners "will be felt as a happy re-creation of their early family lives" (Klein, 1937, p. 74). 'Projective identification' is the subtle behavioral and emotional manipulation of others. Others are *induced* to behave in prescribed ways, "to play a role in the enactment of [our] internal drama – one involving early object relationships" (Cashdan, 1988, p. 56). In the 'projective identification of dependency', the projector displays help-seeking and submissive behaviors and induces care-giving behaviors in the target of the projection. This leads to reenactment of the relationship between infant and primary care giver. If, however, the target of the projection "offers resistance, the projector may experience anxiety, depression, rage" (Cashdan, 1988, p. 77). In the complementary 'projective identification of power', the projector tries to impose a subservient role on the target, "to convince the target that he or she needs to be cared for and looked after" (p. 66). Inducing the target to feel incompetent or inadequate ensures that the projector will be needed and not be abandoned. Thirdly, in the 'projective identification of ingratiation', the projector makes sacrifices, so that the target of the projection feels that he owes something to the projector. A person prone to using this form of projective identification would have learned in childhood that he needs to do things for others and make them feel indebted, so that he will be wanted and appreciated by them (Cashdan, 1988, p. 76).

Neurotic persons, owing to their inner insecurity (basic anxiety), tend to show an excessive dependence on others' approval or affection (Horney, 1937). The neurotic person "reaches out desperately for any kind of affection for the sake of reassurance" (p. 109); he has "an indiscriminate hunger for appreciation or affection" (p. 36). Inner insecurity can manifest as constant requests for others' company, an incapacity to be alone, and a tendency to cling to a significant other (Horney, 1937). Clinging is a 'mode of narcissistic repair' (Bursten, 1973) that is related to proximity-seeking behavior by which infants overcome separation anxiety. 'Craving narcissists' are exceedingly clingy, needy, and demanding in their attitude toward their object; and they act as though they are constantly expecting to be disappointed by their object (Bursten, 1973). Concealed behind the clinging and demanding behavior of craving narcissists, there is a core of grandiosity (the grandiose self), which gives them a sense of entitlement, makes them feel that they have a right to demand attention and care. In craving personalities, compared to other narcissistic personality types, the archaic omnipotent object has been internalized less firmly (Bursten, 1973). 'Oral characters', according to Fenichel (1946), "are fixated on the level of ego development at which their original omnipotence was already lost and they were striving to get it back" (p. 509). They need objects "as pacifiers, protectors, bringers of supplies" (p. 509). Oral characters try to influence their objects by force or ingratiation, to make them furnish narcissistic supplies (Fenichel, 1946).

4.2 Regression to Infantile Dependency

Infantilisms and other forms of explicit care-seeking behavior are readily used by persons of a submissive and dependent character disposition. Infantile gestures and vocalizations *regain* prominence when the personality as a whole *regresses* to an infantile mode of relating to an external figure who has been maneuvered into the role of the primary object. Regression is a 'dependency-seeking process' (Laughlin, 1970, p. 338); it reinstates developmentally earlier 'infantile dependence' (Winnicott, 1989). The ego retreats to a position of greater dependence and safety, "a more protected and less exposed position", which also tends to be a passive position (Laughlin, 1970, p. 322). Regression, as a personality defense, is a "response to overwhelming unconscious needs for safety and security" (p. 322), taking place at times of physical or emotional illness or in response to overwhelming stress. Retreating from "danger, anxiety, stress, and responsibility", the ego "stabilizes itself on an earlier, simpler level" (p. 322) (representing a safer era). Regression presupposes a 'weakness of ego organization' (Laughlin, 1970, p. 334). Narcissistic personalities, even though they may have achieved a degree of internalization of their omnipotent object (affording them with an internal source of approval and an internal mode of self-esteem regulation), may regress to a state of craving (with clinging and demanding behaviors) at times of physical illness or when being drunk (Bursten, 1973). Nevertheless, we all have the potential to regress to an infantile mode of relating, if stresses are sufficiently high and if the environment enables such regression. Regression to an infantile mode of relating can occur if the environment meets our increased dependency needs, that is, if a new 'environmental provision' allows for greater dependency (Winnicott, 1989).

Regression may be associated with psychogenic recapitulation of symptoms that were associated with a serious or prolonged physical illness in early life. Regression to a childhood period of illness occurs "because of the extra care, attention, and love received at such times", especially if parental acceptance or love were lacking at other times (Laughlin, 1970, pp. 331-332). Children and adults, "having once tasted the concern their relatives show for their health", may "use their illness to hijack their family's attention" (Adler, 1927, p. 164). The neurotic person may seek to obtain care and affection by appealing to pity, "by a dramatic demonstration of his complaints", or "by involving himself in a disastrous situation which compels our assistance" (Horney, 1937, p. 141). Not only can anxiety find some resolution through the personality defense of regression, symptoms of anxiety states previously experienced in childhood may be *used* to stabilize the regression. Some persons use anxiety as a "device to compel someone to be close to them and take care of them" (p. 192). Those who strive for superiority as a way of restoring security will feel anxious whenever their sense of superiority is threatened; "[t]heir scream of anxiety" serves to link them "with the world again", in that it ensures that "someone hurries up to them", so that "[t]hey are no longer alone and forgotten" (Adler, 1927, p. 191). Similarly, expressions of grief may be used by the grief-stricken person for the purpose of influencing the attitude of those around him (p. 216). He finds his situation "alleviated by the way others hurry to help [him], sympathize with [him], support [him], encourage [him]" (p. 216). For the same reason, childish 'bad habits' may be maintained into adulthood (p. 201). These habits are "directed towards gaining attention from the adult world"; but unconsciously they aim to ensure that the mother pays constant attention (Adler, 1927, p. 201).

4.3 Illness Behavior

Any illness, physical or mental, can provide 'secondary gain' for the ill person, in that symptoms and illness-related behaviors attract others' attention, sympathy, and affection (Laughlin, 1970, p. 342). Illness is frequently perceived as a right to privileges and bestows economic advantages on the ego "by provoking pity, attention, love, the granting of narcissistic supplies" (Fenichel, 1946, p. 461). Thus, secondary gain, which "consists in getting attention by being sick" (p. 461), can be a factor in maintaining illness. Illness behavior would manifest a "longing for the time of childhood when one was taken care of" (p. 461), a "regression to childhood times, when one was still protected" (p. 462). What is often reactivated at times of illness is a "need of a sign of parental affection and of assurance against abandonment" (p. 461). The expression of this need through illness "may in turn arouse guilt feelings, creating secondary conflicts and vicious circles" (Fenichel, 1946, p. 461).

Intolerable (egodystonic) impulses or ideas are consciously denied but can be *converted* into neurological symptoms. 'Conversion symptoms' allow "the return to consciousness of elements of the conflict ... in a converted and disguised form" (p. 51); they "allow some measure of disguised external expression" and hence partial gratification of disowned impulses (Laughlin, 1970, p. 31). In addition, conversion symptoms provide secondary gain in terms of the fulfilment of dependency needs. Thus, neurological symptoms such as paralysis of a limb, loss of the ability to speak, or functional loss of vision, express intolerable impulses in a consciously more tolerable form and, at the same time, *secure a position of dependency* (p. 43). Similarly, in somatoform disorders, hostility can be converted into headaches and other somatic symptoms, whereby emotional pain (resulting from narcissistic injury) is converted into the experience of bodily pain. Somatoform disorders, such as fatigue states or neurasthenia, can, at the same time, be manifestations of regression (Laughlin, 1970). The tendency to show symptoms of somatization is associated with alexithymia. Alexithymia constitutes "a paucity of affective description and intrapsychic awareness", "a deficiency in usage of affective words" (p. 214), and a predilection to refer to own body parts impersonally and to use "pronouns such as "it", "one", or even "you" ... instead of "I"" (Rickles, 1986, p. 215). Persons with alexithymia relate to others "in rather rigid, emotionally stunted, nonempathic ways"; their capacity for empathic experiencing of objects is lacking (p. 215). Alexithymia predisposes to somatization (psychosomatic breakdown) in response to personal slights, humiliations, and disappointments. Somatization in response to narcissistic injury expresses *demands for dependency* (psychophysiological regression) (p. 219). Persons who are prone to develop psychosomatic disorders have a deficient capacity for self-care and self-soothing; they remain dependent "on others or selfobjects to carry out soothing and self-esteem regulating functions" (Rickles, 1986, p. 218).

The neurotic person is prone to experiencing narcissistic injuries, as "he advances only on condition of being successful in everything he attempts" (Adler, 1938, p. 124). Having been defeated in his striving for superiority (and having thus glimpsed his worthlessness), the person may experience a "fear of being unmasked in all [his] worthlessness" (p. 125), and he may remain in a state of 'shock'. He is "standing before a deep abyss and [is] afraid of being pushed into it – i.e. afraid that [his] worthlessness is going to be revealed" (p. 125). Facing the possibility that his worthlessness (inferiority) will be fully exposed, he "cannot be induced to take a single step forwards" (p. 125). Adler (1938) thought that, by intensifying the physical and psychical 'shock symptoms', the neurotic person can save himself from the full collapse of his self-esteem (p. 130). In his neurosis, the person unknowingly exploits symptoms resulting from the effects of a shock, whereby it "is more feasible for those who have a great dread of losing prestige" to exploit their shock symptoms in this way (p. 135). Neurotic persons "prefer their present suffering to the greater ones they would experience were they to appear defeated" and were they to have their worthlessness revealed (pp. 124-125). Neurotic persons hold on to their symptoms and neurotic suffering because "they are more afraid of something else: of being proved worthless"; they are afraid that "the sinister secret – the fact that they are worthless – might come to light" (Adler, 1938, p. 125).

4.4 Guilt and Self-Punishment

The neurotic person believes that he can be appreciated or loved only for his performances, which means that he has to outperform others who he unconsciously believes compete with him for access to narcissistic supplies (i.e., for the attention of the unconscious omnipotent object). The neurotic person's pride "demands that he *should* be superior to everybody and everything" (Horney, 1950, p. 134). Witnessing others' superior skills or shining qualities "must call forth a self-destructive berating" (p. 134). Competitive attainment of narcissistic supplies is mediated by actualization of the idealized self. Attempting to actualize his idealized self, the neurotic person is bound to fail and, therefore, to hate his actual being (for not being able to "make himself over into something he is not") (p. 374). Frustration in the pursuit of narcissistic resources arouses anger, which, if it cannot be expressed, is rechanneled against the self. The more a person idealizes himself (and the more he is thus alienated from his real self), the more he hates and despises himself for being as he is (Horney, 1950, p. 373). Hatred of the 'real self', arising whenever "we are driven to reach beyond ourselves" (p. 114), manifests in "relentless demands on self, merciless self-accusations, self-contempt, self-frustrations, self-tormenting, and self-destruction" (p. 117). Tormenting himself in imagination, the neurotic person engages "in endless and inconclusive inner dialogues, in which [he] tries to defend himself against his own self-accusations" (p. 145). In self-tormenting, the person "is always both the torturer and the tortured", and "he derives satisfaction from being degraded as well as from degrading himself" (Horney, 1950, p. 148). Self-recriminations are a 'face-saving device' (defense mechanism), in that they provide the neurotic person with "reassurance that he is not so bad after all and that his very qualms of conscience make him better than others" (Horney, 1939, p. 245). When expressed openly, self-recriminations are an attempt to prevent others from making recriminations; they are a strategy of warding off reproaches; they are an attempt to appease others and to elicit their *reassurances* (Horney, 1939, pp. 240-241).

The child, having been punished by his parents for disobedient behavior, shows remorse and asks for forgiveness, in the hope that his parents will take him back into their loving care (Rado, 1956, p. 226). Parental punishment is experienced as a prerequisite for forgiveness. We learn to anticipate relief when we have been punished, to expect that, once we have been punished by our parents, "their anger abates, and they tend to regard the incident as closed" (Flugel, 1945, p. 145). Insofar as "punishment may be a means of achieving forgiveness", "a need for punishment actually may develop" (Fenichel, 1946, p. 138). As Fenichel (1946) pointed out, "the pain of punishment is accepted or even provoked in the hope that after the punishment" the object's affection will be forthcoming (p. 103). Guilt is an expectation or fear of punishment by the superego (or by one of its external representatives). Guilt promotes remorse and confession; it prompts reparative and submissive behavior that appeases the superego (the parental introject) or its external representative. Guilt "elicits the reparative pattern of expiatory behavior, a pattern often complicated by rage retroflexed and vented on the self" (Rado, 1956, p. 245). Developmentally, the child proceeds from expecting parental punishment to punishing himself "for the sake of forgiveness, and the recapturing of the love of the parents, which it entails" (Rado, 1956, p. 226). When there is no external authority figure to inflict the punishment, punishment is attributed to, or unconsciously inflicted by, the 'harsh superego' (Freud, 1923). Self-punishment is essentially punishment of the ego by the superego. Self-punishment terminates the feeling of guilt insofar as it opens the prospect of forgiveness and renewed acceptance by the superego (or by an external representative of the superego).

In this regard, "the relations between the ego and the super-ego mirror those between the child and parent" (Flugel, 1945, p. 145). Thus, the need to be punished and the tendency to punish oneself are crucial components of guilt. 'Retroflexed rage' (against the self) can become "once again environment-directed and defiant" when the self "believes it is acting purely in "self-defense""; the self then "turns from self-reproach to reproaching the very person he guiltily fears" (Rado, 1956, p. 226).

Self-punishment expresses the 'need for punishment' (associated with guilt); it also stems from frustrations in the pursuit of self-actualization (approximation of the ego ideal) and the subsequent rechanneling of aggression against a socially less inacceptable and hence less egodystonic target (the self). Guilt and the need for punishment arise when we breach the standards of the superego, standards enshrined in our ego ideal (as preconditions for acceptance by the superego), whereby such breaches arise from irresistible temptations to gratify desires or enact drive impulses (such as aggressive impulses) *inappropriately*. The need for punishment is inversely correlated with our sense of safety. If we feel reasonably secure, we "have little need for punishment and are free to react to frustration vigorously and aggressively" (Flugel, 1945, p. 163). If we have a degree of insecurity, "we may be liable to feel guilty whenever we enjoy ourselves or express aggression, unless our guilt has been temporarily assuaged by punishment" (p. 163). Self-punishment and suffering can be a form of bribery or tribute paid to the superego, so that "the super-ego permits a certain amount of gratification of desire of which it does not approve" (p. 158). If we feel a "still greater degree of moral insecurity, any expression of impulse, or even temptation of such expression, may arouse overwhelming guilt, and we feel safe only when we suffer; in this case the need for punishment may be almost insatiable" (Flugel, 1945, p. 163). We may provoke disasters or hardship for ourselves in order to satisfy the need for punishment. The tendency to search for and create situations in which punishment can be incurred may be a pervasive personality feature, referred to as 'moral' or 'psychic' masochism (Flugel, 1945; Bergler, 1949, 1952). Those with moral masochism readily seek out a position of victimhood in order to appease their unconscious sense of guilt (Kernberg, 1992). 'Unconscious guilt' ubiquitously manifests in "a proneness to minor self-defeating behaviors", in "characterological inhibitions and self-imposed restrictions of a full enjoyment of life", or in a "tendency for realistic self-criticism to expand into a general depressive mood" (Kernberg, 1992, p. 36). The unconscious sense of guilt can express itself in criminal actions, which allow for a fastening of this guilt "on to something real and immediate" (Freud, 1923, p. 52). Another mechanism is the *projection* of guilt "on to some other party, who is then regarded as responsible, so that we ourselves feel innocent" (and worthy of others' or the superego's approval and acceptance) (Flugel, 1945, p. 150). The need for punishment can be satisfied vicariously, satisfied, that is, by the punishment inflicted on others (p. 174). Projecting our own guilt upon others and witnessing their punishment rebalances our 'moral equilibrium', helping us to preserve our own virtue and self-satisfaction (pp. 168-174). Alternatively, we may suffer somatic or conversion symptoms of neurosis in order to satisfy the need for punishment (Flugel, 1945, p. 155).

4.5 Masochism

Displays of helplessness, weakness, and suffering foster dependent relationships. The masochistic person shows helplessness and "exaggerates his misery and needs for the strategic purpose of getting what he wants" (Horney, 1939, p. 261). He uses suffering and helplessness as means of exerting control and obtaining affection and help (pp. 263-264). Wanting others' attention and care, he submerges himself in feelings of helplessness, unhappiness, and unworthiness; he "exaggerates his weakness and he tenaciously insists on being weak" (p. 268). Moreover, his suffering and helplessness allow him to express "accusations against others in a disguised but effective way" and "to evade all demands others might make on him" (Horney, 1937, p. 264). The masochistic person recoils from self-assertion and is afraid of success; his unobtrusiveness facilitates dependency on others (Horney, 1939). For the masochistic person, total submission to a cruel and powerful object is the only condition for survival (Kernberg, 1992, p. 49). The masochistic person transforms deep feelings of insecurity into a feeling of being in the power of others (Horney, 1937, p. 267). He alleviates his anxiety by thrusting himself on someone's mercy, "[b]y submerging his own individuality entirely and by merging with the partner" (Horney, 1939, p. 253). His security depends on "being absolutely subjected to another's domination" and on *relinquishing the self* (Horney, 1937, p. 274). Submergence in misery produces "satisfaction by losing the self in something greater, by dissolving the individuality, by getting rid of the self with its doubts" (Horney, 1937, p. 270). The masochistic person derives pleasure from exhibiting his misery, because, in spite of the pain and suffering involved, masochistic behavior heightens the feeling of safety (Sandler & Joffe, 1968, p. 262).

The ego of masochistic persons is subjected to excessive superego pressures (Kernberg, 1992). The harsh aspect of the superego ('daimonion' [Bergler]) punishes the ego whenever the latter does not live up to its ideal (ego ideal). The ego counteracts the superego's cruelty by engaging in psychic (moral) masochism. Psychic masochism finds expression in the need for punishment or a need for self-humiliation. The punitive superego of masochistic persons "reflects an unconscious need to suffer as an expiation for guilt feelings" (Kernberg, 1996, p. 126). The masochistic person seeks humiliation and failure in order to be forgiven and accepted by the severe superego (Fenichel, 1946, pp. 363-364). Being submissive and making himself feel small, insignificant, helpless, and pitiful, the masochistic person appeases his superego (as he previously would have appeased his parents). The superego then turns to him (to his ego) with pitying acceptance (and the narcissistic homeostasis is restored). The superego, having been manipulated into a forgiving or pitying stance, softens its demands on the ego. What is enacted here is an internal object relation, the relation between the ego and an internal representative of the primary maternal object. The need to enact an abusive internal object relation may reflect the person's *dependence* on (attachment to) "the "bad mother" who *refuses*" (Bergler, 1949, p. 199).

Psychic masochism can manifest as 'craving for injustice' ('injustice collecting') (Bergler, 1949, 1952). The masochistic person unconsciously provokes the anger of his object in order to feel unjustly treated and victimized. Forgiveness and acceptance back into the object's (superego's) loving care are more likely if the punishment received can be portrayed, in fantasy or reality, as excessive and unwarranted, as an injustice (Bergler, 1949,1952). The masochistic person obtains "narcissistic gratification from the sense of being unjustly treated and thus implicitly morally superior to the object" (Kernberg, 1992, p. 48). The sense of being unjustly treated enables withdrawal into self-pity. The self-pitying person views himself as an innocent victim. Self-pity, in general, is a self-soothing, self-comforting response to a narcissistic injury (Wilson, 1985). Self-pity is derived from the soothing response of the mother to the distressed infant appealing for help. The narcissistic injury that brings about self-pity is typically inflicted by a needed selfobject; it is caused by failure of the selfobject to show empathy (Wilson, 1985). Self-pity allows the narcissistically injured person to recover selfobject connectedness by accessing "a new form of empathy ... that of the self taking itself as its own selfobject" (pp. 187-188). The soothing and comforting object is experienced within the self. Self-pity has "a component of self-righteousness that registers as a complaint against the person who has triggered the reaction"

(Wilson, 1985, p. 188). Instead of pitying himself, the person who feels unjustly treated may react with righteous indignation, "the typical fury of the unjustly treated person" (Bergler, 1949, p. 6). Righteous indignation may alternate with masochistic self-pity. 'Injustice collectors' have a tendency to provoke or imagine a situation in which somebody was unjust and refusing to them, so that they can either pity themselves or revel "in fury and anger from their arsenal of "righteous indignation", seemingly in self-defense" (Bergler, 1949, p. 7).

Through masochistic behavior, which tends to have an accusing and blackmailing tone, the masochistic person attempts to *force* his objects to provide care and give love and affection (Fenichel, 1946, p. 363). Sacrificing and humiliating himself, he tries "to enforce [his] privileges and protection from the omnipotent persons" (Fenichel, 1946, p. 365). The masochistic person puts others under a 'stringent moral obligation' (p. 261); "he is or feels humiliated and suppressed, and in his heart he makes others responsible for his suffering" (Horney, 1939, p. 263). He reacts with resentment and hostility to disappointments, "to the slightest sign of disregard or neglect" (p. 259); and he is bound to become disappointed and resentful, because of his excessive expectations of his objects (Horney, 1939, p. 271). Masochistic persons feel rejected and mistreated in reaction to relatively minor slights, which may lead them to develop "unconscious behaviors geared to making the objects of their love feel guilty" (Kernberg, 1992, p. 37). If 'masochistic trends' are "combined with an imperative need for power and control", then "the masochistic person exerts control ... by his very suffering and helplessness" (Horney, 1939, p. 268). He may "live in a helpless dependence and at the same time exert a tyranny over others by means of his weakness" (Horney, 1937, p. 277). Others "may submit to his wishes because they are afraid that if they do not there will be an upheaval of some kind" (despair or functional disorders) (Horney, 1939, p. 268). Sooner or later however "persons around him become tired of this type of entreaty", "take his misery for granted", and "are no longer spurred to action" (Horney, 1939, p. 261).

The neurotic person with 'inverted sadism' has sadistic impulses but "leans over backward to keep them from being revealed to himself or others" (Horney, 1945, p. 211). "He will shun everything that resembles assertion, aggression, or hostility and as a result will be profoundly and diffusely inhibited" (p. 211). The person with inverted sadism is overanxious not to disappoint others; he will "avoid anything that could conceivably hurt their feelings or in any way humiliate them" (p. 212). Even though he thinks he is fond of people, "he has very little feeling for them at all"; he harbors "unconscious contempt for others, superficially attributed to their lower moral standards" (Horney, 1945, p. 214). The inverted sadist may put up with sadistic behavior directed at himself (p. 214). He may even put himself in the way of exploitation and indulge in feeling victimized, as he seeks and relishes "an opportunity to live out his own sadistic impulses through someone else", whilst, at the same time, feeling innocent and morally indignant (p. 215). Developmentally, the inverted sadist, having been crushed into submission in his childhood, initially developed 'compliant trends' and later in life took refuge in an attitude of 'detachment'; but, as "his need for affection became so desperate", "he became hopeless and developed sadistic trends", which however, as "his need for people was so insistent", had to be repressed and inverted (Horney, 1945, p. 213).

4.6 Depression

Freud (1917) observed that a person who is in mourning tends to feel shame in front of other people, whereas the melancholic patient shows "insistent communicativeness which finds satisfaction in self-exposure" (p. 247). The melancholic patient portrays himself as morally despicable and abases himself before others; he "vilifies himself and expects to be cast out and punished". While he consciously depreciates himself, he *unconsciously* punishes the object he *unconsciously* believes to have lost; the object with whom the ego had become *identified* (Freud, 1917). Persons who are prone to melancholic depression tend to choose their objects on a narcissistic basis, using them predominantly as selfobjects, that is, for the purpose of regulating their self-esteem. Those prone to melancholia are often highly narcissistic and unable to endure injuries to their self-esteem (Nunberg, 1955, p. 143). Narcissistic persons treat their objects as part of their self; they are liable to use *identification* (of the self with an object) as a way of relating to objects. Predilection for narcissistic object choice is associated with ambivalence toward objects, because objects who are *needed but not genuinely loved* have to be controlled, aggressively if necessary. Hostile feelings toward the object, whose care and attention the melancholic person unconsciously believes to have lost, are transformed into punishment of the ego-identified-with-the-object, that is, into self-punishment. Those prone to melancholia also have an excessively severe superego; and self-punishment or self-depreciation in melancholic patients can be attributed to the severe superego. The severe superego can be said to direct its aggression "against the formerly loved person, with whom the ego of the melancholic has become one through identification" (Nunberg, 1955, p. 143).

Having an excessively severe superego also means that the person has an unattainably high ego ideal. Melancholia can also be regarded as an extreme case of 'self-dissatisfaction' brought about by failure to live up to the standards of the ego ideal (Flugel, 1945). The person's ego ideal is high and unattainable (the demands of his ideal are exaggerated), inasmuch as his object relations are loose, that is, inasmuch as he is prone to choosing his objects on a narcissistic basis (Nunberg, 1955, p. 144). Failure to attain the ego ideal (actualize the idealized self) results in loss of love from the superego; and the loss that the melancholic person unconsciously mourns may be this very loss of love from the superego, that is, loss of the reassuring presence of the internal representative of the primary maternal object. Self-punishment (the hallmark of melancholia) then aims to recapture the love of the superego. Nunberg (1955) noticed that melancholia can "set in after professional disappointments or after the patient had had to give up an aim to which he had devoted his entire life" (p. 144). Excessive concern with status, prestige, or possessions is associated with vulnerability to depression. Failures, such as loss of prestige or loss of money, amount to loss of narcissistic supplies, "which the patient had hoped would secure or even enhance his self-esteem" (Fenichel, 1946, p. 390). Loss of prestige or possessions results in "a reduction in the person's apparent loveworthiness" (experienced as sense of worthlessness), which, in turn, is what leads to depression (Flugel, 1945, p. 95). The loss of prestige or possessions unconsciously represents a loss of love from the superego.

It can also be said that the person who is at risk of becoming depressed is "fixated on the state where his self-esteem is regulated by external supplies" (Fenichel, 1946, p. 387). Thus, the loss of self-esteem associated with depression can be due to a loss of internal supplies from the superego or due to a loss of external narcissistic supplies (p. 391). However, the overdependence of depressively disposed persons on their superego merely symbolizes their overdependence on external narcissistic supplies, on sources that act as external superego projections. The key is that persons predisposed to melancholia are unable to love actively; they passively need to be loved (Fenichel, 1946, p. 387). This is what defines their narcissism. They need the feeling of being loved; and without it they "lose their very existence" (Fenichel, 1946, p. 391). Persons predisposed to depression feel secure only when they are loved, esteemed, and supported by others; "their self-esteem largely depends on whether they do or do not meet with approbation and recognition" from their environment (Rado, 1928, p. 49). Persons vulnerable to depression "find difficulty in personal relations because they are ultimately looking for something which they should have had in infancy from their mothers" (Storr, 1968, p. 80). They have not, in their infancy, had their mirroring and idealizing needs met consistently, so that infantile *grandiosity* and exhibitionism persist. Concealed grandiosity in adulthood is related, as Miller (1979) argued, to excessive dependence on others' admiration, representing persistent primitive needs for mirroring and empathy by the maternal object. Concealed grandiosity is a defense against depression; and depression ensues when grandiosity breaks down (Miller, 1979). Concealed grandiosity, emerging intermittently into conscious fantasy, attracts narcissistic supplies from the largerly unconscious omnipotent object (the superego), but efforts have to be made to validate this grandiosity by provoking the selfobject surround into admiring responses. The illusion of grandiosity has to be maintained by continuous outstanding achievements or excessive perfectionism, if the illusion of continued availability and responsiveness of the primary selfobject (the omnipotent object) is to be salvaged (Miller, 1979).

The depressively disposed person is in a relationship of dependency with the 'dominant other', who "is symbolic of the mother, as she appeared to [him] in very early childhood" (Arieti, 1973, p. 129). He "is always afraid of losing or of having already lost what [he] needs from this dominant figure: love, affection, praise, admiration, approval, a precious supply of intangible things that only [she] can give" (p. 129). When failing to meet expectations of the dominant other, the person feels deprived of narcissistic supplies (love, affection, admiration). Depression arises when self-esteem cannot be maintained, because expectations of the dominant other (expectations that were internalized in childhood as parental ideals) cannot be lived up to (Arieti, 1973). The dominant other is a projection (an external replica) of the superego; and the ego ideal or 'idealized image' is the counterpart of a demanding superego. *Hopelessness*, which arises when the person persistently fails to measure up to the idealized image and "becomes aware that he is far from being the uniquely perfect person he sees in his imagination" (p. 184), "is the deeper source from which the depressions emanate" (Horney, 1945, p. 188). The person suffering from 'blaming depression', in particular, is concerned about meeting internalized expectations of the dominant other (effectively the standards of the superego) and feels guilty and *blames* himself for not having met these expectations (p. 129), whereas the person with 'claiming depression' needs an external object from whom to extract praise and approval (Arieti, 1973, p. 130). 'Blaming depression' can give way to 'claiming depression'. If the depressed person becomes fearful about being abandoned by the superego, then his self-esteem diminishes to a danger point, and he makes, in relation to the external world, "desperate attempts to force an object to give the vitally necessary supplies" (Fenichel, 1946, p. 389). When severely depressed, the person draws others "into a relationship in which they are asked to provide things that only a mother could reasonably be asked to provide", that is, to take on basic care-taking functions (Cashdan, 1988, p. 63). By soliciting others' care, the person seeks to restore his narcissistic balance. Many depressive attitudes are formed by condensation of submissiveness and ingratiation, on the one hand, and aggressive control, on the other (Fenichel, 1946, p. 387). Patients with depression often "try to captivate their objects in a way characteristic for masochistic characters, by demonstrating their misery and by accusing the objects of having brought about this misery, and by enforcing and even blackmailing their objects for affection" (Fenichel, 1946, p. 391).

To recapitulate, Freud (1917) thought that those prone to melancholia chose their object on a narcissistic basis. The object is not loved in its own right but controlled (ambivalently) for the purpose of providing love. The object is, in other words, no more than a selfobject; and when control over it is lost, aggressive impulses normally used for the purpose of control are redirected against the self. The same can be said, in a more abstract sense, about the selfobject surround at large; aggression is employed to control this surround but, when control is lost, aggression turns against the self. The self or ego has to be treated as a dynamic concept, as advocated by Federn (1952); the self or ego encroaches onto the external world to varying degrees. When narcissistic control is lost (i.e., when 'narcissistic cathexis' is 'withdrawn' from the external object world), the 'ego boundary', in Federn's words, retracts; it shifts from the external to the internal world. This is somewhat equivalent to stating, with Freud (1917), that in melancholia object cathexis is abandoned and regresses to narcissistic identification with the object: the ego identifies with the object, and the ego-identified-with-the-object thus becomes the target of aggressive impulses previously used to control the object. At the same time, aggression is attributed to the superego, that is, to the 'critical agency' that is split off from the ego (Freud, 1917, p. 248). The superego turns against the ego-identified-with-the-object "with the same rage that this ego previously used in its struggle with the object" (Fenichel, 1946, p. 393). Self-reproaches, a key feature of depression, may shift back to reproaches and hostility directed against the 'lost' object (insofar as the loss concerns only a loss of control) or against a substitute object. The patient comes to believe (unconsciously) "that the object alone was to blame" (Rado, 1928, p. 55). Thereafter, it may be the ego again that "is thrust into the place of the hated object" (p. 56), so that "the super-ego visits upon the ego all the fury which the ego would otherwise have been capable of visiting upon the object" (Rado, 1928, p. 55).

Self-reproaches, as Money-Kyrle (1961) recognized, may be a derivative of the aggression used by the infant for the purpose of controlling the responsiveness of, and the care and attention received from, the primary object (pp. 20-21). Self-reproaches and self-punishment also express remorse and seek forgiveness from the superego, the internal representative of the primary object. The depressed patient "begs for forgiveness and endeavours in this way to win back the lost object" or, if he cannot procure "the pardon and love of his object, he tries to secure those of his super-ego" (Rado, 1928, p. 50). The patient's struggle for the love of his object, his narcissistic desire to be loved, is thus carried over to his relation with his superego (p. 50). What takes place in melancholia is a "regression from an object-relation to a narcissistic substitute for it" (p. 47), "a narcissistic flight from the object-relation to that with the super-ego" (p. 50). The ego, in the state of depression, depreciates itself in order to get assurances of love from the superego. Replicating the cycle of loss of love from the parents and subsequent parental punishment and forgiveness, the ego punishes itself and seeks expiation from the superego, that is, from their internal mental representative (p. 51). In melancholia, self-punishment takes place in the hope of absolution, of final forgiveness and acceptance (Rado, 1928, p. 51). The patient's ego "submits to the cruelest torments, to the very point of self-destruction, in order thus to regain the blissful situation of being loved" (Rado, 1956, p. 46).

4.7 Summary

Appeasement gestures and displays of submissiveness are related to infantile care-seeking behaviors; and both types of behavior condense in the infant's display of helplessness. Submissive gestures, on the one hand, and care-seeking behaviors (including infantile displays of helplessness), on the other, elicit related forms of narcissistic nourishment, namely acceptance and affectionateness, respectively, which are attitudes others adopt toward oneself when their aggressiveness is inhibited. Expressions of helplessness, weakness, and suffering are means not only of obtaining narcissistic supplies but also of controlling the relationship to the object (selfobject) for the purpose of retaining one's safety. Predilection to narcissistic object choice and overdependence on narcissistic supplies from external representatives of the superego render oneself vulnerable to depression. Failure to control the (narcissistically used) object or selfobject surround provokes angry attempts to *reassert control*; and, if these attempts remain chronically frustrated, depression ensues, wherein aggression is rechanneled against the self (in the hope of expiation from the superego, or in a last effort to force the self into an *acceptable* shape).

Frustration in the pursuit of narcissistic resources or failure to meet demands and expectations of the superego (enshrined in the ego ideal of idealized self) causes anger, which (in order to avoid further disapproval by the superego) may have to be rechanneled against the self, manifesting as self-recriminations. When engaging in self-recriminations, the self (ego) derives some pleasure from degrading itself, which is probably related to an expectation of reinstated safety. The need for punishment of the self, associated with conscious or unconscious guilt, expresses an expectation of forgiveness and renewed acceptance by the superego. Similarly, self-recriminations or self-defeating behaviors may be a way of appeasing the superego in order to allow oneself to gratify instinctual desires of which the superego would normally disapprove. Psychic masochism involves feelings of being victimized or unjustly treated as well as behaviors that aim to rekindle these feelings in oneself (injustice collection). The masochist can feel self-righteous and morally superior and accuse the object of neglect, thereby inducing feelings of guilt in and tightening control over the object. At the same time or alternatively, feelings of being victimized or treated unjustly enable withdrawal into the state of self-pity, wherein the (inner) self receives soothing responses from the introjected object (the superego).

The external and internal world are in some sense mirror images of each other; and their relationship to the self is essentially that of the mother to the infant. Hostility of external powers toward oneself (aspects of oneself embedded in the external world) can be perceived, unconsciously, as prelude to receiving their forgiveness, and thus, if these *external* powers are superego projections, the forgiveness of the superego. Alternatively, the hostility of external agents and one's victimhood can be a pretext for self-pity (the self, here, as part of the imaginary world) and for seeking pity and consolation from yet higher powers (for one's being treated unjustly), namely from *inner* and more ephemeral representatives of the superego. External powers are provoked, unconsciously, to engage in punishing actions, actions that can then be perceived as unjust. Unjust aggression from external powers fosters one's own sense of virtuousness, which is ultimately one's sense of being acceptable and close to the superego in the inner world. Finally, self-punishment, which lies at the core of depression, may be an expression of rage against the object, insofar as the object has become identified with the ego (and ego and superego would have become dedifferentiated) in a final attempt to restore the narcissistic balance (bypassing secondary narcissism and reverting to primary narcissism).

Chapter 5

Detachment

The neurotic person, as Horney (1937) argued, "is excessively afraid of or hypersensitive to being disapproved of, criticized, accused, found out" (p. 235); "he cannot help believing that others will despise him ... if they find out about his weaknesses" (p. 240). Being criticized by others or being defeated in competition means for the neurotic person that he has to realize a definite weakness or shortcoming, a realization that for him is unbearable. Hypersensitivity to being disapproved of or criticized is associated with a feeling of 'intrinsic weakness', that is, "a deep feeling of insignificance or rather of nothingness", which, in turn, is covered up by a grandiose but fragile façade or 'persona' (p. 267). The neurotic person has "high-flung notions of his uniqueness" (p. 265); yet these are easily shattered (Horney, 1937). Kohut (1966) suggested that infantile grandiosity (the grandiosity of the 'narcissistic self') persists into adulthood when it serves to cover over a precarious self. If infantile grandiosity is insufficiently modified during personality development, the person remains liable to be narcissistically injured and to experience shame, because ambitions are likely to be thwarted when they are based on irrational overestimation of the self (Kohut, 1966). When, in Adlerian terms, the 'ideal goal of perfection', which is the 'guiding fiction' in the pursuit of superiority, is out of reach and "too unrealistic to constitute an incentive to real endeavour", then "the person takes refuge in phantasies, in an attempt to regain in this way his sense of worth" (Flugel, 1945, p. 47). Flight into fantasy requires an attitude of detachment, which, in itself, is a defense against the fear of being disapproved of and thus forced to face up to one's inferiority. Hypersensitivity to being disapproved of or criticized can result in a complete withdrawal of all feelings (emotional investments), manifesting as emotional unresponsiveness and coldness (Horney, 1937, p. 136). The person has "to become emotionally detached from people so that nothing will hurt or disappoint" (Horney, 1937, p. 99).

Characterological detachment is an active movement away from relationships and a freezing of the capacity for relatedness (Schecter, 1978). Characterological detachment resembles detachment experienced by infants in response to separation from the mother (active avoidance of the mother) (following initial stages of 'angry protest' and 'despair') (Bowlby, 1973). Transitory detachment is part of the process of mourning in adults. Characterological detachment is a stable system of defenses that protects against psychic pain arising from abandonment or loss of love (Schecter, 1978). Persons with milder forms of detachment have shallow relationships. Their fear of dependency and "a great dread of the death of the loved person" (p. 86) lead to repeated rejections of the object ("or to stifling and denying love") and to "the repeated turning away from a (loved) object", while being "driven from one person to another" (infidelity) (Klein, 1937, p. 86). Schizoid persons maintain a particularly severe and enduring detachment. Schizoid persons keep themselves "at a safe emotional distance from others" (via emotional coldness), which protects them "against the danger of incurring a permanent or protracted breakup, enfeeblement, or serious distortion of the self" (Kohut, 1977, p. 192).

Characterological detachment is accompanied by periodic regression to fusion with internal objects (symbiotic fantasies), aiming to fill the extreme isolation of the detached person (Schecter, 1978, p. 83). Detachment is related to the adaptive state of 'aloneness'. Aloneness, as opposed to loneliness, presupposes the presence in psychic reality of another, an internalized good object. The infant's developing capacity to be 'alone in the presence of the mother' is a precursor of the mature capacity to be alone and all by oneself, that is, to seek out and enjoy a relationship solely with oneself (Winnicott, 1958). The subjective self, being a derivative of (and remaining inherently connected to) objective others, is never alone, as Winnicott (1958) pointed out. Inner life, being played out in a state of aloneness, features the self in relationship with good internal objects (although these may be hidden from conscious awareness); and it is through these internal object relations that the self takes shape and becomes cohesive.

5.1 Omnipotence

During the developmental phase of 'separation-individuation' (from about five to 24 months of age), the child becomes aware of his limitations and of his dependence on the parents (while also developing pleasure in independent functioning) (Mahler, 1967, 1972). As a result of nontraumatic frustrations and threats of object loss, the child comes to realize his separateness. At the same time, becoming aware of his limitations, his self is gradually divested from omnipotence and overestimation. He must gradually and painfully give up the delusion of his grandeur (Mahler, 1967, 1972). His self-esteem may undergo critical deflation, however identification with an emotionally available mother and internalization of a good child-mother relationship help to restore self-esteem. Nevertheless, the child continues to long for the lost 'ideal state of the self' (Joffe & Sandler, 1965), in which he was *merged with the mother*; and this longing continues to be a spur for development. After his 'psychological birth' (recognition of his separateness), the child enters a life-long struggle against isolation, involving efforts to return to or approximate the 'ideal state of self', the state of fusion with the symbiotic mother (Mahler, 1967, 1972).

Relinquishing the belief in his own omnipotence, the child tries, at first, to participate in the perceived omnipotence of his parents. Having had to renounce his belief in his omnipotence, the child "considers the adults who have now become independent objects to be omnipotent, and he tries by introjection to share their omnipotence again" (Fenichel, 1946, p. 40). The child participates in their omnipotence whenever he feels loved by them. Narcissistic feelings of wellbeing, revived in this way, "are felt as a reunion with an omnipotent force in the external world" (Fenichel, 1946, p. 40). When the child has to realize that he can no longer participate in his parents' omnipotence, he develops an increased sense of dependency and increasingly suffers separation anxiety; and this process is in balance with his resurging strivings for separateness and autonomy. During the 'rapprochement' subphase of the separation-individuation phase (second half of the second year), the child alternates between demanding or protecting his autonomy and demanding or seeking his mother's closeness (Mahler, 1972). Mahler (1972) suggested that "the entire life cycle [constitutes] a more or less successful process of distancing from and introjection of the lost symbiotic mother, an eternal longing for the actual or fantasized 'ideal state of self', with the latter standing for a symbiotic fusion with the 'all good' symbiotic mother, who was at one time part of the self in a blissful state of well-being" (safety).

Feelings of omnipotence are gradually modified in development (becoming realistic goals), if the child experiences narcissistic injuries and frustrations of his demands in the context of maternal empathy and understanding (Rothstein, 1979). Approximation of the 'ideal state of self' does not generally entail restoration of infantile omnipotence, provided that the primary maternal object has been introjected successfully and a good internal object relation has supplanted an external one. This is somewhat equivalent to stating that the state of primary narcissism is approached indirectly by way of secondary narcissism; that safety (Sandler) is attained through narcissistically gratifying relationships with external *and* *internal* objects. The omnipotent child, on the other hand, is self-sufficient and independent in his fantasy; he turns the 'negative realization' of the 'no breast' into a feeling that he does not need the object after all (Bion, 1962). Assignment of omnipotence back to the self (after omnipotence had previously been overcome in development, first by assignment of omnipotence to the parents) is defensive in nature. If the child's capacity to tolerate frustration is low, then he may use omnipotence as a means for tolerating frustration (Bion, 1962).

5.2 Hidden Grandiosity

The detached person, as has already been pointed out, is estranged from others; he avoids close and lasting relationships with them (Horney, 1945). He has to maintain his detachment in order to avoid rejection and narcissistic injury, but his detachment enables him, at the same time, to hide intolerable feelings of inferiority under notions of his own greatness. The detached person sees himself as a superior or unique being, as someone who is bound to attract others' approval or admiration without having to reach out to and engage with them. The detached person also harbors "fantasies of a future when he would accomplish exceptional things" (Horney, 1945, p. 79). Feelings and fantasies of superiority or uniqueness allow the detached person to uphold his self-esteem, yet these feelings and fantasies flourish only in a state of detachment. As Horney (1945) stated, the detached person's remoteness is "an over-all protection to which he must tenaciously cling and which he must defend at whatever cost" (p. 92). There are various behaviors and attitudes that the person develops for the purpose of maintaining his detachment and thereby protecting notions of his own greatness from crumbling or being exposed to ridicule. He may think of himself as an independent or independently minded person and act in accordance with this notion. The need for independence, when it amounts to characterological detachment, "manifests itself in a hypersensitivity to everything in any way resembling coercion, influence, obligation" (Horney, 1945, p. 77). The detached person avoids anything that could jeopardize his detachment. Long-term obligations are avoided; "the thought of joining any movement or professional group where real participation and not merely payment of dues is required" causes feelings of panic (p. 88). Others' expectations toward the detached person, even timetables, constitute a threat and make him feel uneasy and rebellious (p. 78). "He will conform outwardly in order to avoid friction, but in his own mind he stubbornly rejects all conventional rules and standards" (Horney, 1945, p. 78).

The idealized self-image of neurotic persons is grandiose and unrealistic and hence associated with proneness to detachment; and both the idealized image and detachment provide the characterological ground for indecisiveness and an inability to take responsibility. Neurotic persons tend to be unable to take responsibility for themselves, to acknowledge to themselves and to others their intentions, and to accept the consequences of their actions (Horney, 1945). The neurotic person, when faced with the consequences of his actions, "often tries to wiggle out by denying, forgetting, belittling" or placing responsibility on others (p. 171). Recognizing the consequences of his actions and assuming responsibility for them would shatter his hidden feeling of omnipotence (p. 172). His idealized image "does not permit of the possibility of being wrong", so "he must falsify matters and ascribe the adverse consequences to someone else" (Horney, 1945, p. 173). Similarly, neurotic persons "often cannot commit themselves to a feeling or opinion about another person"; they often cannot take "a stand in accordance with the objective merits of a person, idea, or cause"; however, they "are readily swayed – unconsciously bribed, as it were – by the lure of greater affection, greater prestige, recognition, power" (pp. 168-169). They rationalize as fairness their inability to make up their mind or take a stand, whereby such fairness can in itself be a compulsory part of the idealized image (Horney, 1945, p. 169).

Even though the detached person keeps himself at a safe distance from others, he can release and express a variety of feelings, including creative feelings, as long as he expresses them in a manner that is not directly connected with human relationships (Horney, 1945, p. 83). For the neurotic person, who lacks spontaneity and has pervasive inhibitions in areas of affectionateness and assertiveness, "detachment will provide the best chance of expressing what creative ability there is" (p. 90). Detachment promotes intellectualization. Inasmuch as emotions have to be checked, "emphasis will be placed upon intelligence", with an expectation that "everything can be solved by sheer power of reasoning" (Horney, 1945, p. 85). In schizoid persons, whose character-ological detachment is particularly severe, preoccupation with the inner world coexists with 'overvaluation of thought' and a tendency to substitute intellectual for practical solutions of their emotional problems (Fairbairn, 1952, p. 20). Schizoid persons "are often more inclined to construct intellectual systems of an elaborate kind than to develop emotional relationships with others on a human basis" (p. 21). They circumvent emotional difficulties by focusing their efforts on attainments in the intellectual realm (Fairbairn, 1952, p. 23). Schizoid persons concentrate "their often considerable libidinal resources on pursuits which minimize human contact (such as interest and work in the area of aesthetics; or the study of abstract, theoretical topics)" (Kohut, 1971, p. 12). Detachment, by facilitating intellectualization, reinforces concealed grandiosity. Schizoid persons are prone "to look down from their intellectual retreats upon common humanity with a superior attitude" (Fairbairn, 1952, p. 21). They have a sense of inner superiority, even though this is often largely unconscious (p. 22). Their attitude of superiority may be "concealed under a superficial attitude of inferiority or humility; and it may be consciously cherished as a precious secret" (Fairbairn, 1952, p. 7).

5.3 Grandiose Fantasy

Approval, acceptance, and love can be sought not only externally from others (in the social context) but also from the internal world, by withdrawing into the private realm of fantasy (Laughlin, 1970). Daydreams often have a narcissistic emphasis; events and interactions in daydreams, featuring the daydreamer himself, provide narcissistic gratification (Federn, 1952; Sandler & Nagera, 1963). Daydreaming can be a pathological form of self-esteem regulation when it "indulges in timeless sham events, actually substituting for the present" (Federn, 1952, p. 360). Fantasy allows for self-absorption and self-glorification and is conducive to character-ological detachment. Fantasy is a potent means for avoiding conflict and avoiding realistic action (Laughlin, 1970, p. 119).

The child may perceive his mother to be unreliable and become profoundly disillusioned with her early on. 'Precocious separation' from the mother leads to a precocious internalization of the self, that is, to the formation of a fragile sense of self that needs to be supported by fantasies of grandiosity and omnipotence (Modell, 1975). At the same time, the child may develop a fear of closeness, which manifests in characterological detachment, which facilitates indulgence in and withdrawal into grandiose or omnipotent fantasies (Modell, 1975). Grandiose fantasies about oneself reveal a grandiose ego ideal. The ego ideal of narcissistic persons, in particular, has grandiose proportions (Kernberg, 1970). This would make it difficult for these persons to actualize their ideal and to maintain their narcissistic balance other than in grandiose fantasy. Thus, narcissistic persons may withdraw from social life and into fantasy as effectively as those with schizoid personality disorder (Kernberg, 1970). It can be argued that the schizoid character is another form of narcissistic personality disturbance, one in which omnipotence and grandiosity are combined with hypersensitivity to rejection (or heightened fear of disapproval) and lack of assertiveness. Grandiose fantasies are the predominant mode of expression of the grandiose ego ideal, that is, of the ego's craving for acceptance by (and fusion with) the omnipotent object (superego), if the person is highly sensitive to rejection and fearful of closeness, and if he thus prefers to operate in a mode of detachment.

There is perhaps a slight distinction to be made between omnipotent and grandiose fantasies. Omnipotent fantasies concern illusions of self-sufficiency and independence from the primary object, whereas grandiose fantasies, insofar as they can be distinguished from omnipotent fantasies, express the need for acceptance by a concrete or abstract or unconscious representative of the primary object. The more or less conscious notion of one's grandiosity counteracts deep-seated feelings of worthlessness and reestablishes the person's worth or acceptability in the eyes of a concrete or abstract derivative of the primary maternal object, whereby this derivative can be conscious or unconscious, with conscious derivatives comprising internal and external projections or incarnations of the superego or omnipotent object. Fantasies featuring a grandiose self are ultimately fantasies about an internal omnipotent object, whether the object features in these fantasies or not. Narcissistic persons aim to achieve reunion with their omnipotent object; and they do so by embellishing their self in fantasy, if not by living up to their ego ideal in reality. When interacting with reality, shortcomings and weaknesses of the personality become apparent. Shortcomings and weaknesses, as judged against the values that have been internalized as the ego ideal, are a source of shame and a threat to the objective of reunion with the omnipotent object (Bursten, 1973). Shortcoming and weaknesses are easily overlooked in fantasy.

5.4 Faith in God and Belief in Fate

God is a superego projection par excellence. Our relationship with God is that of our ego with the superego (Flugel, 1945). God provides narcissistic sustenance in the form of higher recognition and divine approval. God is a 'supernatural provider' who "must be appeased by prayer and self-torture" (Erikson, 1950, p. 225), much as the superego has to be appeased by guilt and self-recrimination. Religions "abound with efforts at atonement which try to make up for vague deeds against a maternal matrix" (Erikson, 1950, p. 225). God ultimately represents our striving for acceptance by and reunion with the primary maternal object. Adler (1938) suggested that "the idea of God ... corresponds to the obscure yearning of human beings to reach perfection" (p. 199). To reach perfection and become acceptable to God, we have to live up to the ideal prescribed by our religious culture, which is internalized as the ego ideal. The ego ideal is, as Freud (1923) saw it, "a substitute for a longing for the father"; and as such "it contains the germ from which all religions have evolved" (p. 37). The patriarchal leader lost his significance when, in the course of cultural evolution, ancient kinship structures disintegrated and social structures became more fluid. With the demise of the patriarchal leader, who himself was a developmental derivative of the primary maternal object, God became a substitute anchor for safety and compass for social conduct.

Gods are commonly portrayed as helpful and benign beings as well as punishing and malignant beings (Flugel, 1945, p. 187). God has loving and protecting as well as frustrating and punishing aspects, much as our parents had, whose introjection gave rise to the superego, the origin of God. The ambivalence in our attitude to God is a reflection of the ambivalence in our attitude to our parents, who once acted as sources of approval and as sources of disapproval and punishment. When religions try to reconcile "the all-powerfulness and all-lovingness of God" with the existence of evil (p. 187), they pick up on our often unconscious 'need for punishment'. Temptation by a forbidden impulse or commitment of a sin arouses in us the need to subject ourselves to punishment, so that we can regain the good will and love of God, much as we had earlier in our life tried to recapture the love of our parents by accepting and enduring their punishment (Flugel, 1945). Fate is another, but "more shadowy" projection of the superego and ultimately also a representation of our parents (Flugel, 1945, p. 161). Being delivered a blow by fate is equivalent to the withdrawal of parental love and protection. "When fate delivers us a blow", "we may be overcome by the same sense of loneliness and helplessness that we experienced in early years if we imagined our parents had forsaken us" (p. 161). When, however, "fate 'smiles' upon us", we feel once more as though we are "'basking' in the approval of the parents" (Flugel, 1945, p. 161).

Gambling, according to Fenichel (1946), "is a provocation of fate, which is forced to make its decision for or against the individual" (p. 372). Luck in gambling "means a promise of protection (and of narcissistic supplies)" (Fenichel, 1946, p. 372). Greenson (1947) argued that the neurotic gambler has strong yearnings for omnipotence and that, by way of gambling, he "seeks a sign from Fate", a sign that would confirm his omnipotence (p. 7). His gambling challenges fate or God "to calm his grave doubts" about his omnipotence (p. 7). Being lucky means to the neurotic gambler that he is omnipotent. Alternatively, or at the same time, being lucky means that the gambler has been accepted by God, so that he can share in God's omnipotence (p. 8). Winning evokes euphoria (feelings of triumph) "because it represents reunion with the omnipotent one" (Greenson, 1947, p. 10). Winning, in other words, revives a derivative of the 'oceanic feeling' experienced by the infant in a state of union with the mother (Fenichel, 1946, p. 39). Greenson (1947) thought that "the neurotic gambler has regressed to infantile longings for omnipotence" and that "strong longings for omnipotence and oceanic feelings are evidence of a failure of the ego to maintain a mature level" (p. 9). Losing, by contrast, means being abandoned by fate or God and can lead to depression. Greenson (1947) pointed out that the neurotic gambler, who "is a personality on the brink of a severe depression" (p. 8), has actually a need to lose, despite the depression which losing can cause. The need to lose is an expression of unconscious guilt, of an unconscious need for punishment with its promise of absolution or forgiveness. Punishment in the form of losing "is a lesser evil than the terrifying punishment of castration or total loss of love" (Greenson, 1947, p. 11).

5.5 Schizoid Personality Disorder

Schizoid persons have an attitude of detachment and social isolation; they keep their objects away from themselves. They cannot express their feelings naturally toward others, finding it difficult to act naturally and spontaneously (Fairbairn, 1952, p. 20). Persons with schizophrenia, too, "had shown some signs of peculiarity from early on in life and were never able to express strong feelings"; they have always had "a tendency to turn away from the outside world at the least provocation" (Rosenfeld, 1965, p. 167). Schizoid persons have "learned to distance themselves from others in order to avoid the specific danger of exposing themselves to a narcissistic injury" (p. 12); their distancing is a consequence of "the correct assessment of their narcissistic vulnerability" (Kohut, 1971, p. 12). Schizoid persons are unable to endure narcissistic injuries (p. 532) and readily react to frustrations with 'partial loss of object cathexes' and withdrawal into state of 'primary and omnipotent narcissism', a state in which they see themselves as independent from others (Fenichel, 1946, p. 531). Schizoid persons became convinced early in life that "their mother did not really love them and value them" (Fairbairn, 1952, p. 23). Their mother failed "to convince her child by spontaneous and genuine expressions of affection that she herself loves him as a person" (Fairbairn, 1952, p. 13).

The schizoid person "longs deep down to love and be loved", but "he can only permit himself to love and be loved from afar off" (Fairbairn, 1952, p. 26). The schizoid person has a longing to be seen, known, and recognized by others, hoping thereby to preserve his fragile identity (Laing, 1960).

His fear of loss of his self, that is, his "fear of being invisible, of disappearing, is closely associated with the fear of his mother disappearing" (p. 125). At the same time, being seen by others or being just visible to them poses a threat to his identity and his sense of realness (i.e., to the cohesiveness of his self). The schizoid person "feels more 'vulnerable', more liable to be exposed by the look of another person" (p. 79). While there is a longing to be seen or known, to be seen or known is also what is most dreaded (p. 123). The schizoid person longs for a moment of recognition (p. 122), however "of this very longing he is terrified" (Laing, 1960, p. 97). Giving in to this longing means facing the possibility of ultimate rejection or abandonment. Not only schizoid persons, but also those with a schizophrenic condition, experience a "perpetual conflict between the wish for human contact" and "the danger of excessive closeness" (Reiser, 1986, p. 232). Due to their fear of rejection, schizoid and schizophrenic persons are unable to use selfobjects to maintain their narcissistic balance; "they are forever at risk and forever alone", however "sooner or later the dreadful loneliness catches up and drives the schizophrenic patient to seek human contact, despite the risks" (Reiser, 1986, p. 232). For schizoid persons, participation in life is possible "but only in the face of intense anxiety" (Laing, 1960, p. 95). Insofar as contact can be made, schizoid persons treat their objects "as means of satisfying their own requirements rather than as persons possessing inherent value" (Fairbairn, 1952, p. 13) (i.e., as selfobjects). Objects that are 'incorporated into the self' but also perceived to be part of the external world (selfobjects) have to be controlled compulsively, if the person is unable to form mature relationships (if he has not progressed to 'mature dependence'). The person who is fixated on an infantile attitude of 'taking' (characteristic of the stage of 'infantile dependence') resorts to exhibitionism, masochism, or sadism (Fairbairn, 1952) in his efforts to compulsively control his selfobjects ('incorporated objects'). The exclusively narcissistic use of objects is characteristic especially for narcissistic personality and behavior disorders, which emphasizes again the continuity between schizoid and narcissistic conditions.

Schizoid persons are preoccupied with their inner reality and overvalue thought processes (i.e., they intellectualize). They had to transfer, early in life, their relationship with external objects into "the realm of inner reality"; "their objects tend to belong to the inner rather than to the outer world" (Fairbairn, 1952, p. 18). In other words, schizoid persons deal with their lack of gratifying external object relationships by turning to internalized objects. Inasmuch as interest ('libido') is withdrawn from external objects, it is directed toward internalized objects, so that it is in the "inner reality that the values of the schizoid individual are to be found" (p. 50). Schizoid persons build up the 'libidinal' (or rather 'narcissistic') value of their objects in the inner world; and "they tend to identify themselves very strongly with their internal objects" (Fairbairn, 1952, p. 18). Identification with internalized objects is coupled with a sense of *secrete* possession of these objects, which results in a narcissistic inflation of the ego, a *secret* sense of superiority (p. 22). The necessity of secrecy is partly "determined by fear of the loss of internalized objects which appear infinitely precious (even as precious as life itself)" (p. 22). Secret possession of internalized objects causes the person "to feel that he is 'different' from other people – even if not, as often happens, actually exceptional or unique" (Fairbairn, 1952, p. 22). The sense of difference from others (uniqueness) is also a consequence of the need to offset feelings of inferiority toward objects in the outer world. However, the attitude of superiority, adopted by schizoid persons for defensive purposes, "is based upon an orientation towards internalized objects" (Fairbairn, 1952, pp. 50-51). By contrast, the superiority of those with narcissistic personality disorder, as commonly understood, is based on narcissistic supplies demanded and drawn from *external* objects.

Schizoid fantasies feature internalized objects with which the self identifies, a grandiose self that is elevated to the level of the unconsciously fantasized omnipotent object, or an omnipotent self that is independent and does not need any objects. As the schizoid person withdraws into fantasy, he constructs and maintains an 'inner self', which he separates from his 'false self' or persona, the latter being the 'identity-for-others' which arises in compliance with others' expectations (Laing, 1960, p. 105). The inner self, occupying the realm of fantasy, is omnipotent; it "can be anyone, anywhere, do anything, have everything" (p. 88). The illusion of omnipotence and freedom can be sustained only in fantasy. To protect the inner self against the danger of destruction from outer sources, the schizoid person has to eliminate "any direct access from without to this 'inner' self" (p. 152). However, "what was designed in the first instance as a guard or barrier to prevent disruptive impingements on the self, can become the walls of a prison from which the self cannot escape" (Laing, 1960, p. 150).

5.6 Psychosis

Kohut (1977) thought that schizophrenia is a form of pathology of the 'fragmented self' (p. 243) that arises in consequence of the emotional distance of selfobjects in childhood (early selfobject failures) (p. 257). Federn (1952) previously argued that 'weakness of the ego' is a fundamental problem in schizophrenia and psychosis (pp. 105, 166). He, too, thought that mothers of psychotic patients tend to be afflicted with 'strong narcissism', which opposes the 'devotion without hesitation' that is part of the 'instinctual behavior pattern' of normal motherhood (Federn, 1952, pp. 144-145). Adverse experiences in early life, such as maternal rejection, are often repressed but can be reactivated later in life, such as when the person comes to feel "that the segment of the world that is important to him finds him unacceptable" (Arieti, 1973, p. 126). The person who is at risk of developing schizophrenia then "sees himself as totally defeated, without any worth and possibility of redemption" (p. 127) and realizes "that as long as he lives he will be unacceptable to others" (p. 126). He may then adopt new forms of cognition, allowing him to transform "the intrapsychic danger into an external or interpersonal one" (Arieti, 1973, p. 127). Once psychosis has established itself, aggressiveness, which was previously suppressed, gets closer to the threshold of expression, being aimed against those who frustrate the person's longing for human relatedness and who have a competitive advantage in the struggle for narcissistic resources. Instead of "a desperate *longing* and yearning for what others have and he lacks", he experiences "frantic *envy* and hatred of all that is theirs and not his, or a desire to destroy all the goodness, freshness, richness in the world" (Laing 1960, p. 96). Such hatred may not be translated into aggressive actions at first, that is, for as long as an egosyntonic outlet for aggression has not presented itself.

Psychosis involves severe detachment, to a point at which the external world is experienced as dreamlike, unreal, or dead (derealization). The ego, so to say, withdraws narcissistic cathexes from a world that refuses to meet its narcissistic needs. Patients may even experience themselves or their thinking as automatic and mechanical (Schilder, 1976, p. 43). Such depersonalization (the counterpart of derealization) involves excessive self-observation. Through self-observation the ego sets itself above its narcissistic needs. Withdrawal from the world is also coupled with "an incipient tendency to destroy the world and oneself" (Schilder, 1976, p. 43), which would be limited at first to hatred and a desire to destroy. The patient, who "no longer dares to fully experience his own body and the world", gives them up but "retains his full interest in merely observing his incapacity", adopting, as 'the destroyer' of the world and of himself, the stance of a '*supreme* observer' (Schilder, 1976, p. 43). Decisions cannot be made as they would draw the person back into the external world. Patients with catatonic schizophrenia, in particular, find themselves immobilized when faced with a decision. Catatonic immobilization or stupor "seems specifically to reflect a radically exaggerated obsessional hesitancy, indecision, and precautionary concern" (Shapiro, 2000, p. 149).

In a state of isolation and narcissistic aloofness, the psychotic patient can "overcompensate all narcissistic hurts by developing a still higher opinion of himself" (Fenichel, 1946, p. 421). In narcissistic daydreams, he experiences himself as being more wonderful than anyone else (Fenichel, 1946, p. 421). Grandiose fantasies of psychotic and hence severely detached patients are related to the idealized image of which Horney spoke. When grandiose fantasies start to be corroborated by subjective (including *hallucinatory*) experience, grandiose delusions develop (and these can stabilize in a delusional system). Patients may not only have grandiose ideas about themselves, they may also, as their psychosis develops, feel and act as though they are a person of great significance. Megalomania is an extreme form of pathological overvaluation of the self. The megalomanic patient, "who proclaims that he is Lincoln, Christ, or Napoleon", "actually *feels* himself to be greatest and most important of human specimens" (Hendrick, 1958, p. 119). The patient would, at the same time, act in accordance with demands and expectations of reality, a seeming contradiction that is known as 'double orientation'.

Detachment is allied with denial of reality, or at least of part thereof. Denial is a primitive defense against conflict and anxiety and forms, along with projection, the basis of persecutory delusions (Laughlin, 1970). Projection of hostile impulses operates in most of the paranoid and persecutory delusions. The patient wishes to harm his objects but cannot do so for fear of retaliation, hence he expects to be harmed by them instead (Laughlin, 1970, pp. 230-232). It is this expectation of being harmed by others that allows for the projection of self-contempt. Feeling defeated and blaming himself, the patient externalizes (projects) his self-blame and self-accusation, so that, instead of accusing himself, he feels that the accusation comes from the external world (Arieti, 1973, p. 127). This gives him a new focus for his aggression (a potentially egosyntonic one). Moreover, when the "accused person now is not the patient but the persecutor" ("who is accused of persecuting the patient"), "the patient experiences a rise in his self-esteem, often accompanied by a feeling of martyrdom", given especially that "the patient feels falsely accused" (Arieti, 1973, p. 127). The persecuted patient sees himself as a person of great consequence when he has become "the subject of widespread attention, even if this be malicious" (Storr, 1968, p. 95). In acting "as though some sinister deity is singling [him] out for persecution" and as though "all hostile powers are intent on wreaking vengeance upon [him]", the patient reveals his vanity, his need to feel important (Adler, 1927, p. 210). Persecutory delusions are economically beneficial to the ego, as manifested in the phenomenon of 'paranoid indifference', which patients show when reporting their persecution (Laughlin, 1970, p. 234).

5.7 Summary

The sense of self is a sense of separateness and, at the same time, a longing for connectedness and for reattainment of the primary narcissistic state of fusion with the object (whereby, in a state of fusion, the self dissolves). The secure self is bound (unconsciously) to good internal objects (particularly the superego, the introjected maternal object) or their external replicas. Insecurity of the self (corresponding to a lack of securely established internal objects) and sensitivity to rejection lead to compensatory grandiosity of the (inner) self and to detachment (whereby detachment is maintained both to prevent narcissistic injuries and to foster feelings of grandiosity). The insecure (and precocious) self tends to be omnipotent and grandiose and to fluctuate between denial of dependence and dependence on an omnipotent internal object. Omnipotence is a state both of independence from objects and of longing for fusion with an omnipotent object (superego). This longing expresses itself in grandiose fantasies or in ambitions guided by a grandiose ego ideal. Fantasies featuring a grandiose self are usually not communicated because others' reactions would produce shame, further explaining the need for detachment. The inner self (reflecting others' views of oneself), too, would feel shame, so grandiose fantasies are often tolerated only at the margins of consciousness. There is a balance, struck differently in different personalities, between the tendency to preoccupy oneself with grandiose fantasies (in a state of detachment) and the more or less compulsive pursuit of self-realization (realization of the ego ideal, which is the goal of the personality conjured up in imagery), a balance that depends on the person's intensity of narcissistic need (degree of insecurity), his sensitivity to narcissistic injury, and his endowment with social and physical attributes and adaptive coping skills.

A sense of self arises in interaction with superego projections embedded in the external world. The 'false self' or persona, which the person (not only the neurotic person) maintains in compliance with the external world, is a special aspect of this self. The self that is orientated toward the external world, including the false self, needs to be distinguished from the 'inner self'. It is the inner self that features in grandiose fantasies and that demands and receives sustenance from internalized objects. The inner self, being sustained in and by the inner world, has to be insulated from the outer world. The inner self lives in a state of fantasy, whereas the ego ideal briefly features in imagery and incentivizes reality-oriented pursuits. On the one hand, grandiose fantasies fulfill a protective function, preventing disintegration anxieties or shame concerning an insufficiently secure self. On the other hand, excessive engagement in grandiose fantasies further isolates the person and thereby adds to his vulnerability to ridicule and to his inferiority (sense of vulnerability). Although detachment is a source of vulnerability, it has to be held firmly in place by schizoid persons, as it is a precondition for defenses, apart from being a defense in its own right. The neurotic (and especially the schizoid) person maintains his detachment (and thereby protects his omnipotence or grandiosity) by entertaining, and acting in accordance with, views of himself as someone who is particularly independently minded or objective in his dealings with the world. Intellectualization strengthens grandiosity but depends on a degree of detachment. Intellectualization can be adaptive, when employed as part of abstract pursuits, but it can also be used to build delusional systems, which are ultimately concerned with justifying one's grandiosity (as a condition of safety). In psychosis, detachment is carried to the extreme (although even psychosis can be adaptive in the context of modern mental health systems). Not only grandiose but also persecutory delusions (substantiated in part by hallucinatory experiences) enhance self-esteem, but persecutory delusions become the main feature of psychosis if frustration and envy arise and can no longer be contained in self-contempt and self-blame, that is, if self-contempt and self-blame are channeled away from the self and justifiable external targets for aggression are required.

Chapter 6

Idealization and Identification

Children have a propensity to admire and idealize their object (a selfobject) and thereby to merge with it. Idealization of the selfobject flows into identification with it. Early in the child's life, the mother's "holding and carrying allows merger-experiences with the self-object's idealized omnipotence" (Kohut, 1977, p. 179). Merger can take place with an "empathic omnipotent idealized self-object" (p. 85), a role played first by the mother and then the father. The emphatically responsive father has to "allow himself to enjoy being idealized by his son" (p. 12). Idealization causes enhancement of self-esteem "via the temporary participation in the omnipotence of the idealized self-object" (p. 13). The child's "relation to the empathically responding self-object parent who permits and indeed enjoys the child's idealization of him and merger with him" establishes the child's cohesive 'idealized parent imago' (p. 185). The 'idealized parent imago', representing the wish to merge with an idealized selfobject, is one of the two polar areas of the 'nuclear self', the other one being the 'grandiose-exhibitionistic self' (Kohut, 1977, p. 49) ('narcissistic self' [Kohut, 1966]). The child's archaic wish to merge with an omnipotent selfobject is transformed, in the course of development, into 'attainable ideals', which, along with 'realistic ambitions' (derived from the grandiose-exhibitionistic self), underpin capacities to obtain narcissistic sustenance from the realistic selfobject surround and to thereby maintain self-esteem (Kohut, 1977, p. 82). As development proceeds, the child's idealizations (investments of 'idealizing narcissistic libido'), which retain their narcissistic character, will normally coexist and become integrated with object-instinctual cathexes, namely love ('object libido') and hate. 'Idealizing narcissistic libido' (which is invested in selfobjects) "plays a significant role in mature object relationships, where it is amalgamated with true object libido" (Kohut, 1971, p. 40).

The child, when identified with the parent, "feels not only fearless and protected, but also of increased bodily size and freed from the experience of the weakness of his ego" (Federn, 1952, p. 350). By identifying with the idealized parent, the child can harness narcissistic supplies available to that object and thereby enhance his own self-esteem and wellbeing (Sandler, 1960b). When the child, as a result of identification with the object, feels "the same as the admired and idealized object", "some of the libidinal cathexis of the object is transferred to the self"; "the child feels loved and obtains an inner state of well-being" (p. 36). In other words, the child feels *liked* inasmuch as the object is liked and admired; "the esteem in which the omnipotent and admired object is held is duplicated in the self and gives rise to self-esteem" (Sandler, 1960b, p. 36). The child's archaic idealization of his parent (the 'idealized parent imago') is developmentally continuous with idealizations of the parental objects in the late preoedipal period and the oedipal period (Kohut, 1971, p. 40). During the oedipal period, identification with the same-sex parent allows the child to satisfy his desire for attention from the opposite-sex parent. Furthermore, idealization of the parental selfobject allows the child "to acquire (i.e., to integrate into his own self) certain of his father's abilities" (Kohut, 1977, p. 11). Attributes of the idealized object are "attractive and appealing, and thus are far more readily emulated or taken over" (Laughlin, 1970, p. 130). When adopting traits, mannerisms, goals, or attitudes of the idealized object, the child learns to act out "derivatives of his grandiose and exhibitionistic strivings" "in an aim-inhibited, socially acceptable way" (Kohut, 1977, p. 11). Idealization and identification thus importantly contribute to character formation, to the formation of psychic structures that help to regulate the narcissistic balance on increasingly abstract levels. Identification with the parent and, later in life, with a leader or cultural ideal entails enactment of these ideals and leads to the acquisition of parentally and then socially *approved* behaviors. Through identification with the parent or, later in life, with a person of high social standing, the child or adult gains acceptance, recognition, and approval (narcissistic supplies) and hence security (safety) (Laughlin, 1970); "the acquisition of civilized habits bestows a feeling of heightened self-esteem" (Kohut, 1977, p. 112).

In fantasies about oneself, but also in interactions with the social surround, we often play an adopted role. We identify with our ego ideal, which in itself is a product of identifications. Furthermore, we identify with others when observing their interactions. Money-Kyrle (1961) recognized that "our everyday reasoning about our fellows is anthropomorphic and based on identification" (p. 17). Inferences about others' behaviors and intentions rest ultimately on our sense of being partially identified with them (pp. 22-23). The capacity to observe the self, that is, to be self-conscious, is founded on "the power to identify with other people and so to perceive the world, with the self in it, from their point of view" (Money-Kyrle, 1961, p. 78). There is only a small step from observing or monitoring others in the external world while being identified with them to observing the self in its interactions (with imaginary competitors and superego replicas) in internal imagery. In states of introspection, we can be identified with another (such as a successful competitor for the attention of an internal superego projection) and, at the same time, see ourselves through the eyes of yet another (the superego). We enact the ego ideal or idealized self-image (the product of previous identifications) and, at the same time, judge ourselves from a higher perspective, from the stance of the introjected parent.

6.1 Identification in Object Relationships

Identification with another person may be a means of attracting and binding a *third* person, as happens in the Oedipal scenario. By identifying with another person and assuming some characteristics of this person ("because one wishes to be like this person"), one can "potentially receive whatever the other person does" (Schilder, 1951, p. 273). This relates especially to narcissistic supplies attainable from the primary object or from one of its derivatives (superego replicas). Identification is generally "motivated by deep basic needs for acceptance, approval, and love" (p. 135) and by the desire for "acceptance or love from the object" (Laughlin, 1970, p. 146). One can gratify one's exhibitionistic impulses (and realize one's ambitions) by being like the other person and becoming as approvable and praiseworthy (or loveable) as the other person. The idealized person would usually be introjected and integrated into the ego ideal, so that henceforth the identification will be one with the ego ideal. The ego ideal then acts as the attractive and love-worthy person, but what the ego tries to attract is the love of the superego (the introjected primary object). Identification is also part of the more primitive scenario that involves idealization directly of the primary object or of one of its derivatives. Personality development may arrest at the level of admiration of and identification with a concrete person taking the place of the primary object. 'Ideal-hungry personalities' search for selfobjects whose wealth or prestige they can admire or whom they can admire for their power or attractiveness (Kohut & Wolf, 1978). Ideal-hungry personalities need "selfobjects to whom they can look up and by whom they can feel accepted" (Wolf, 1988, p. 73). Ideal-hungry personalities feel worthwhile only if they are in the company of idealizable selfobjects (Kohut & Wolf, 1978).

Instinctual, including exhibitionistic, wishes can be gratified via a proxy, a person with whom one identifies. In a form of altruism, forbidden instinctual wishes are projected onto a person with whom the altruist identifies (A. Freud,

1937). The altruist takes a friendly interest in someone whose instinctual wishes represent his own wishes and then, by way of identification, gratifies his wishes passively through that person. The person with whom the altruist identifies, and who can better fulfil the altruist's wishes, would once have been envied by the altruist. Altruism thus provides a defense against envy (A. Freud, 1937). Wishes that are gratified in this way are often derived from exhibitionistic impulses; they represent the altruist's own ambitions. When identified with the previously envied person, the altruist can gratify his exhibitionistic impulses and ambitions by witnessing the person, who is now subject to altruistic investments, being in receipt of approval and praise. The altruist may use aggression to control the person with whom he identifies; and therein he may find a justified outlet for his aggressive impulses.

'Identification with the aggressor' is another defense mechanism in which identification plays a prominent role (A. Freud, 1937). By identifying with and impersonating a powerful authority figure, and taking on the authority's aggressive attitudes, the subjugated person can bolster his self and defend himself against feelings of helplessness or impotence. The mechanism of identification with the aggressor contributes to superego formation, in that the child identifies with the punishing aspect of the parental authority and thereby internalizes parental criticism (A. Freud, 1937). The child learns to criticize himself, or indeed to criticize another who takes on the role of the self. When the child identifies with the critical aspects of his superego, he can project, at the same time, his self-image onto another (Sandler, 1960b, p. 41). By identifying with the superego introject (the introjected parental authority), the child (or adult) can adopt a critical and moralizing attitude toward another person. This mechanism effects a defense against feelings of guilt associated with unacceptable wishes and explains why "those who most vocally proclaim moral precepts are often those who feel most guilty about their own unconscious wish to do what they criticize in others" (Sandler, 1960b, p. 41).

6.2 Identification with the Group or Leader

Members of a group show heightened suggestibility and readiness for identification with their leader and with each other (Freud, 1921). Identification with the leader flows from admiration and adoration of the leader. The leader of the group is the object of each group member's idealization. At the same time, each group member believes that he is individually loved by the leader (Freud, 1921). The group derives its cohesion in part from the idea, held more or less consciously by each member, that he is individually *loved* by the leader, who is an omnipotent object standing in developmental continuity with the primary maternal object. The leader (like originally the parent) not only allows participation in his omnipotence (via idealization) but also provides narcissistic nourishment in the form of mirroring responses and signs of acceptance. A common 'basic assumption' shared by members of a group is that "the group exists in order to be sustained by a leader on whom it depends for nourishment, material and spiritual, and protection" (Bion, 1952, p. 78). The leader in the basic-assumption group of 'dependency' is a symbol of the protecting and sustaining parent. Members behave as if they are inadequate, immature, and helpless. Their behavior (reminiscent of the infant's care-seeking behavior) is designed to induce the leader to meet their dependency needs (Bion, 1952). In the process of group formation, as Scheidlinger (1968) confirmed, there is commonly an initial dependency phase, involving "a regression of the group members to a dependent state in relation to a leader" (p. 241). During this phase, "group members in a shared fantasy appear to seek nurture and support from a magical parent-leader" (p. 245). Fantasized gratification of the wish for union with the leader generates euphoria and contentment in each group member (Scheidlinger, 1968).

The common tie of group members with their leader allows them to identify with each other and to feel united, to have a group identity (Freud, 1921). Identification of members with the group as a whole counterbalances competition between members for exclusive attention from the leader. Being identified with the group, members overcome their rivalries with each other and their envious attitudes toward each other. Each group member, when identifying with the group, "reacts to the attributes of the group as if these attributes were also his own" (Scheidlinger, 1964, p. 223). Identification with the group entity is associated with idealization of the group. The group or organization to which one belongs "can serve as an idealizable selfobject – a source of pride in belonging to it – and may also provide a self-confirming selfobject experience" (Wolf, 1988, pp. 47-48). The feeling aspect of idealization is a remnant of the primary narcissistic experience. Identification with a group (extension of each member's ego boundaries to a common identification) represents a partial reversal to the state of primary narcissism; that is, "the primary narcissistic cathexis unity ... may be renewed at the occasion of the expansion of the ego boundaries into the group ego" (Federn, 1952, p. 350). According to Scheidlinger (1964), the group, satisfying each member's needs for belonging and protection, is, on a deep level, "the symbolic representation of a nurturing mother" (p. 218). The need for group membership is, on a deep level, "the wish for reunion with a nurturing mother" (p. 229). The need to belong to and identify with a group "represents a covert wish for restoring an earlier state of unconflicted well-being inherent in the exclusive union with mother" (p. 218), "the original state of unconflicted well-being represented in the earliest infant-mother tie" (Scheidlinger, 1964, p. 224). *Identification* (union) with the mother, implying a state of 'unconflicted wellbeing' (safety), developmentally precedes the formation of a *relationship* with the mother, a relationship that is based on recognition of the separateness of the object and the self. Freud (1921) proposed that, in group formation, there takes place a regression from object choice to narcissistic identification. In other words, the relationship between self and object, that is, between self and leader or between ego and superego, reverts back to, or rather approximates, the state of primary narcissistic union.

In a cohesive group, the self becomes obsolete; group members merge their selves within the group. Self, ego ideal, and superego are psychological structures that regulate the person's narcissistic balance outside primitive group processes. The self (ego), forming outside the context of a cohesive group, is a historically more recent safety devise, a modern detour to the state of safety (wellbeing). Ego identity, ego defenses (i.e., mechanisms that preserve the self), and interactions between self and superego require and bind 'energy'. Energy bound in the internal world can be released and passed on to the cohesive group. The satisfaction of breaking through the 'shell of individuality' (p. 274) and losing oneself in something greater is illustrated by the phenomenon of enthusiasm for a cause (and can also be discerned in love and the masochistic attitude) (Horney, 1937, p. 272). Enthusiasm, which is characterized by elation in association with increased activity, "makes the ego's defenses unnecessary", so that "[t]he energy which has been used for defense can now be utilized for enthusiasm" (Greenson, 1962, p. 180). The person filled with enthusiasm is generous and has urgent or compelling "wish to share with others" (p. 172). Enthusiasm is contagious. When enthusiasm spreads, "there is the feeling of joining and being a member of a group – a feeling of belonging" (p. 172). Those who do not share in the group's enthusiasm "often have a feeling of being left out, cheated – of not belonging"; and they may become envious (p. 172). Those who can evoke enthusiasm amongst others often become leaders, the ability to incite enthusiasm being an important characteristic of leaders (Greenson, 1962).

The individual can overcome his limitations and sense of *isolation* by surrendering the self to a common cause, "[b]y dissolving the self in something greater, by becoming part of a greater entity" (Horney, 1937, p. 273). The merging of the individual with the group is accompanied by loss of the sense of individual separateness or distinctiveness (Freud, 1921). Individuals lack distinctiveness especially when they are members of a group that operates in line with one of the 'basic assumptions' (Bion, 1952). When the group differentiates out of a basic-assumptions state, the self emerges, separating itself from the group identity and resuming its distinctiveness. Self-experience would at first be partial and would remain linked to the leader of the group, rather than the superego. The term 'ego identity' (Erikson) perhaps best describes this intermediate state, when the self has already formed but is not yet isolated, when the self is still dependent on the leader, as an external object, but not yet on the superego, as an internal structure. Ego identity, as Erikson emphasized, is a group psychological phenomenon, which implies coexistence of individual identity and identification with the group. Identification with the group entity "promotes an individual sense of belonging, of enhanced self-esteem, and of ego identity" (Scheidlinger, 1964, p. 226). Belonging to the group means "giving up of some aspect of the individual's self" and giving it over to the group (Scheidlinger, 1964, p. 220); it does not necessarily mean complete surrender and dissolution of the self. In principle, however, it can be maintained that the self, as a defensive structure (as a means for restoring the narcissistic homeostasis), dissolves as ego boundaries extend across the group.

6.3 Identification with God

God, as Flugel (1945) explained, represents our parents and our superego (p. 167). The superego, having been formed by introjection of our parents, can be projected out again. We project the superego onto the external figure of God, who "is in some ways the most suitable of all figures for projection of the super-ego" (p. 186). God is "the perfect loving parent" (p. 187), "a divine parent of whose power and infallibility" we can be assured (p. 186). Religion gratifies our "wish for a protecting, kindly, omnipotent, and omniscient parent" (p. 225). We feel safe in the presence of "an omnipotent Creator who watches over us lovingly" (p. 268). In our relationship with God, we can "enjoy a continuation of the protection and guidance that was given to us by our parents in our infancy" (p. 268). Flugel (1945) suggested that "in the idea of God we are able to recapture that sense of reliance on an all-good, all-wise parent which we enjoyed in our early years and which, we had regretfully come to realize, could not be permanently and completely satisfied in reference to any purely human figure" (p. 262). Suffering and helplessness increase the need for God; "men have most need of God when they feel themselves most helpless in the face of evil", even though "the very existence of this evil might seem to belie the divine love" (p. 268). Evil and suffering create "so great a need for superhuman help that men will cling all the more desperately to the belief in such a love" (pp. 268-269). The helplessness that is associated with suffering and hardship "naturally induces a tendency to regression to the infantile position when we were dependent upon our early parents" (Flugel, 1945, p. 269).

The sense of safety provided by God can increase at times and take the form of religious exaltation, much as healthy self-esteem is on a continuum with manic elation. Religious exaltation, as Flugel (1945) recognized, arises "from a sense of unity with the divine, a unity that seems to correspond psychologically to some condition of fusion between the ego and the super-ego" (p. 262). The raising of the worshipper to the level of his God in states of religious exaltation corresponds to "the ego being somehow raised to the position of the superego, the child to that of the parent" (p. 270). In ecstatic religious experience, the ego loses its 'petty individuality' as it merges with the divine representative of the superego. When the worshipper is 'in tune with the infinite', his separation anxiety is completely alleviated, and he can enjoy a feeling of bliss and harmony (Flugel, 1945, p. 186). The sense of safety and wellbeing is transiently restored in the fullest measure; the oceanic feeling associated with primary narcissism is closely approximated.

6.4 Infatuation

The lover experiences exaltation in the presence of his beloved object. The lover is in a state of joyful surrender and experiences an expansion of his personality (Flugel, 1945). The beloved object narcissistically nourishes the lover, similarly to how the child feels safe and joyful in the presence of the loving and caring parent. Flugel (1945) suggested that, in the state of love, the loving aspect of the superego (the derivative of the loving parent) is projected onto the beloved object. Working through the beloved object, the superego acts "to embrace, attract, and elevate the ego" (p. 179). The projected superego has the power "to exercise a sthenic and elevating effect upon the ego, to raise the ego to its own level and there to undergo in some respects a fusion with it" (p. 180). Inhibitions are reduced and conflicts are overcome in the presence of the external figure onto whom the loving and caring aspects of the superego have been projected (Flugel, 1945). Energy that was hitherto invested into the maintenance of the ego is freed. When fusion with the superego has been achieved, energetic efforts aimed at soliciting the approval of the superego and attracting the loving attention of the superego become superfluous. Thus, in the presence of a beloved object (who not only acts as an external representative of the superego but who also gives love unconditionally and is receptive to idealizations), the lover experiences "greater freedom and availability of mental energy" (Flugel, 1945, p. 180).

Patients with borderline personality disorder have an intense hunger for objects. Unlike narcissistic patients, they do not attempt to maintain an illusion of self-sufficiency (Modell, 1975). Borderline patients meet narcissistic needs by idealization of an external object and participation in the omnipotence of the idealized object, whereas narcissistic patients rely predominantly on their grandiose self (as counterpart of the unconscious omnipotent object [Bursten, 1973]), maintaining the grandiose self by self-deception and in fantasy or by forcing mirroring responses from their selfobject milieu. Annie Reich (1953) described female patients who used a primitive form of identification ('primary identification') as a way of relating to their external objects. Idealization of their object was associated with adoption of the object's personality traits and interests (identification with the object). Having idealized and identified with their object, the women described by Annie Reich (1953) were able to share in their object's grandeur, allowing them to feel great and wonderful themselves. These primary identifications (involving superficial imitations) served as a substitute for real object love and real object relations. Annie Reich's patients (1953) readily glorified and identified with anyone whose worth was recognized by other people. Any minor criticism of their object by a third person caused the narcissistic overvaluation of the object to break down and the identification to be relinquished, leading to a sudden drop in self-esteem and hostility toward the abandoned object. These women then turned to a new object, which was again rapidly idealized and identified with. Annie Reich (1953) suggested that such rapid identifications offer narcissistic compensation for narcissistic injuries sustained in childhood and provide a temporary substitute for lacking real object relationships. Very similar rapid alternations between idealization and vilification with associated fluctuations in self-esteem and mood can be seen in patients with borderline personality disorder.

6.5 Summary

Idealization of an object entails identification with the object and participation in its perceived omnipotence. When idealizing an object, the ego merges with the object; the ego becomes identified with it. Primary identification, in the sense of merger with the object, is a primitive process, characteristic of preoedipal stages of development (and of cohesive groups centered around a patriarchal leader). By contrast, identification with an admired or envied *third* object in a triadic (oedipal) constellation strengthens and develops the ego. The ego here emulates (identifies with) the third object (which acts as an ideal) and, in doing so, unconsciously seeks to merge (identify) with the primary object (or a developmental derivative thereof). The ego ideal is the precipitate of such oedipal identifications, as Freud recognized. The ego can identify with its ideal in deed or fantasy. Ego identity (Erikson) describes the ego effortlessly acting in accordance with these identifications, in the knowledge of being assured the approval of the group or leader. When identifying itself with its ideal in *fantasy*, the ego defines itself in form of a subjective (inner) self and positions itself into the view of the superego, giving rise to the phenomenon of self-observation. Self-observation accentuates the separateness of the self but would occur with the aim (unconsciously) of preparing merger with the superego. Ego (self), ego ideal, and superego regulate the person's narcissistic balance outside primitive group processes. The self, as a subjective and introspective phenomenon, is an adaptation to complex and fluid social structures; it aids orientation in social processes that are not clearly centered on a single and unambiguous heir of the primary object. Psychological processes between self and superego, including ego defenses, consume energy, the energy that is released in states of enthusiasm and mania and can be utilized when the cohesive group acts in unison. Outside primitive group processes, energy has to be expended to elevate the self to the level of the superego in the hope of replicating the safety-procuring bond between infant and mother or between group member and leader. Owing to the multilayered, changeable, and uncertain nature of the social world in which we live, the reality-oriented self cannot maintain itself for long at a level at which fusion with any superego replica is possible, but whenever this fusion is briefly achieved, energetic efforts (including ego defenses) aimed at gaining the approval of the superego are superfluous.

Conclusions

Narcissistic sustenance can, firstly, be solicited from the object in the form of recognition, respect, or approval and, secondly, be obtained by participation in the object's omnipotence. These two principle ways of regulating the narcissistic equilibrium correspond to two archaic narcissistic configurations, the narcissistic (grandiose-exhibitionistic) self (seeking mirroring responses from selfobjects) and the idealized parent imago (Kohut). Narcissistic (approval-seeking) behavior involves grandiose and exhibitionistic displays (which may be interwoven with appeasement signals). The object's mirroring responses (approval, praise), actively solicited by the subject's grandiose and exhibitionistic displays, replicate the 'gleam in the mother's eye' (associated with the mother's loving devotion to the infant) and thereby restore the subject's narcissistic equilibrium (Kohut). The second form of narcissism seeks to reinstate the primary narcissistic union with the mother (or with a later representative of the mother) in a more direct manner. When idealizing an object, the underlying unconscious fantasy is that self and object are merged. Idealization of the object entails identification with the object. Gratification of narcissistic needs (in either form) renews the feeling of safety (Sandler), counteracting the danger of being aggressed by the mother or the group (with the potential consequence of annihilation). The discussion shall now be limited to behaviors and psychological mechanisms that seek to generate safety by inviting (or expecting) mirroring (approving, admiring, recognizing) responses from the selfobject surround (including the mother as its earliest representative). The recurrent movement, across the social landscape, from a state of anxiety (signaling danger) to a state of safety, taking into account cultural and situational factors and adjusting time and again to naturally occurring perturbations in the selfobject surround, is what fuels defensive and character structures and imparts on the personality its apparent intentionality and goal-directedness (Adler). Narcissistic behaviors, which in their habitually used constellation characterize a particular personality type, recreate and maintain the self (ego) as an encapsulation of the person's relative safety and of his potential to obtain narcissistic sustenance in an uncertain and inherently dangerous social world.

Narcissistic behavior, that is, the seeking of positive attention (approval and recognition) from others is probably an evolutionary derivative of attachment behavior. Proximity-seeking behavior, that is, attachment behavior in the narrow sense (Bowlby), can be regarded, along with narcissistic behavior (Behrendt, 2015), as an expression of the self-preservative drive (Silverman, 1991; Goodman, 2002). Separation anxiety (Bowlby) would be on a continuum with (and the evolutionary precursor of) the type of anxiety that is arises when one does not receive positive attention from the social surround or when one realizes one's separateness (distinctiveness) from the object (without there being spatial separation from the object) (Rothstein, 1979). This form of anxiety, being probably identical with Kohut's 'disintegration anxiety' and Horney's 'basic anxiety', is counterbalanced by self-experience, representing one's connectedness to the social surround and one's closeness to the superego that unconsciously structures this surround. Ego defenses maintain the integrity of the self (or 'ego'); they can therefore be said to operate in the interest of the self-preservative drive (in accordance with classical psychoanalytic theory). Preservation and integrity of the self mean that the individual is safe in a (mostly latently) hostile social world, which is equivalent to the individual being acknowledged, recognized, and approved by others (as these are attitudes that signal the inhibition of others' aggressiveness). Disintegration anxiety arises when needed narcissistic sustenance is not received (despite being sought) (Kohut) or when ego defenses break down.

Self-preservation can have two meanings, relating to inter- and intraspecific aggression. Firstly, with regards to the need to avoid becoming the victim of interspecific (predatory) aggression, the infant's movements toward the mother and efforts to stay in the proximity of the mother complements the infant's attempts to attract the mother's attention. Separation anxiety is coupled with attention-seeking behavior. The second meaning of self-preservation relates to the inherent aggressiveness of the mother (as recognized by Storr [1968] and others). The infant has to employ behaviors aimed at appeasing the mother. In species with pronounced intraspecific aggression, obtaining the mother's attention would not be enough; her aggressive potential would have to be inhibited, too. Likewise, it is not enough to be in the focus of the group's or the leader's attention; the aggressive potential of conspecifics has to be inhibited, and it is constantly being inhibited by appeasement gestures woven into the fabric of social behavior (Lorenz, 1963; Hass, 1968; Storr, 1968; Eibl-Eibesfeldt, 1970; Moynihan, 1998). The individual, in other words, has to keep paranoid anxiety (which Klein understood forms a substratum of psychic organization) or fears of annihilation (Fenichel) at bay (by way of situationally appropriate narcissistic behaviors, including appeasement gestures). Paranoid anxiety and disintegration anxiety are probably closely related to each other (although the former refers more the external world and the latter more to the self). Paranoid anxiety would be an evolutionary derivative of separation anxiety, much as predatory (interspecific) aggression (to which the infant exposes himself when becoming separated from his mother) was the likely evolutionary predecessor of intraspecific aggression. Developmentally, separation anxiety precedes stranger anxiety, the first manifestation of paranoid or social anxiety. The seeking of others' attention, originally borne out of separation anxiety, remains an integral part of the spectrum of behaviors used to appease others and inhibit their aggressive potential.

Compliance with social norms inhibits others' offensive aggressiveness, because it signals to them acceptance of their social position or rank. Compliance also appeases the superego and safeguards the superego's love, much as compliance appeased the parents and ensured continuation of their loving care. Assertiveness is another method of retaining the parents' or object's love and ensuring their ongoing commitment to oneself. Assertiveness, an aim-inhibited form of intraspecific (offensive) aggression, can also protect access to narcissistic supplies from the wider selfobject surround; it can help to ensure that abstract superego projects continue to provide supplies of approval, respect, and recognition, which are needed to maintain one's safety (vis-à-vis the group's or leader's aggressive potential). Access to narcissistic resources is controlled in a manner that is not dissimilar to the way in which territorial boundaries are protected. Territorial aggression is a form of intraspecific aggression; but intraspecific aggression can also be used to protect one's ranking position in the social order (abstractly, one's proximity to the representative of the primary object), which more clearly defines one's access to narcissistic resources and one's safety in an environment of latent or overt mutual aggressiveness. Assertive control of the object, for the sake of ensuring the object's commitment, can spill over into overt aggression against the object. In a relationship, aggression can induce submissive (respectful) behavior (which provides a form of narcissistic sustenance) in the object and bind the object to oneself more tightly (through its aversive and punishing effects on the object), thereby maintaining the context in which safety can be experienced. The principle of subordinating others for the purpose of attaining safety is starkly illustrated by sadistic attitudes and behaviors. Masochistic attitudes and behaviors, too, have as their aim the binding of the object to oneself, again for the sake of approximating the sense of safety that was first experienced in the state of primary narcissism (the undifferentiated union of mother and infant). Sadistic and masochistic behaviors serve the purpose of maintaining the object's *availability* and *responsiveness*. Submission to others or conformity with norms as well as assertion of dominance or subordination of others generate a safe context for the expression of exhibitionistic and affectionate impulses, which are more directly concerned with the solicitation of narcissistic sustenance.

In a relationship, partners unconsciously assign roles to each other and induce each other to respond in certain ways, replicating patterns of interaction established in childhood. These 'role relationships' are a vehicle for the attainment of safety (Sandler & Sandler, 1978). The way in which the individual relates to the group is a reflection of early object relations, too (Scheidlinger, 1964, 1968). Social situations are generated in daily life with the objective, unconsciously, of attaining or preserving the goodwill and responsiveness of a projected version of the superego (especially the 'dominant other' [Arieti, 1973]). Interactions within a group are competitive and collaborative (pursuing a common goal defined by the leader) and serve to confirm or challenge hierarchical (dominance) patterns and alliances, all of which define the individual's proximity to the leader (or dominant other) or his acceptance by the group as a whole (either of which is a representative of the primary object). The way in which the social environment at large is perceived and shaped is continuous with the infant's attempts to overcome the anxiety associated with the realization of his separateness from the object and the fear of the object's potential aggressiveness (Klein's 'paranoid-schizoid' developmental position) (as well as the anxiety associated with realizing the dependence on the object ['depressive position']). The social environment is patterned by the projected superego (the representative of the primary object) and perceived with reference to the individual's concerns about his safety (which is guaranteed by the projected superego but is also under threat from the projected superego and the group). It is also from the social environment, that the individual extricates his sense of self;

the environment is perceived as a set of references to the self. While the self (ego) reflects the individual's connectedness to the social surround, that is, his acceptance and potential to be approved by others (especially the 'dominant other' and other superego projects) as well as the effectiveness of his attitudes and behaviors geared toward inhibiting others' aggressive potential (ultimately the aggressiveness of the primary object and hence the superego), the ego ideal (ideal self) relates to the individual's *desired* state of safety. The ego ideal is constituted and reshaped by way of imitating successful persons encountered in the course of development, persons who are attractive for and readily approved by the dominant other. The individual identifies with role models and emulates his ego ideal in order to please authority figures and thus to feel safe.

The self as an *internal* image of oneself depends on approbation received from imaginary objects (representatives of internal objects). The self as an internal image is visible to and therefore approvable by an imaginary audience (Cashdan, 1988) (which is usually not consciously elaborated). When thinking about oneself, one intermittently adopts someone else's perspective. By virtue of this identification with an other, the self is looking at itself; the self is an object to itself (Federn). The superego represents this audience and this observing self. The observing self is, in other words, the self identified with the superego, which is also the *inner* representative of the primary object. Feeling the need to be accepted and approved by the superego, one adopts the perspective of the superego, so as to consider from this perspective one's (the ego's) worthiness of approval. In a state of detachment, when conscious fantasy is prolonged, the self can transform itself into its ideal (self-glorification) and thus reach the height of the superego (and potentially reunite with it) without the need to engage with the external social world. The ego ideal, when the ego identifies with it in conscious fantasy, entails an expectation of approval from the superego (narcissistic expectation). For this reason, the ego ideal can act as an incentive goal for behavior concerned with enhancing one's approvability (in the eyes of external superego projects). The ego ideal sets a goal to be realized by ambitions, the derivative of the infant's exhibitionism (Kohut). The ideal self (conscious instatement of the ego ideal, according to Sandler, if the ego ideal were to be regarded as an unconscious construct) guides exhibitionistic and ambitious behaviors aimed at reengaging the leader (or another superego project). Other forms of self-imagery (Horney's 'idealized image') serve similar ends. Imagery of a contemptuous and guilty self leads to behavior that invites punishment from the leader (or any other superego project) with the objective of establishing the leader's forgiveness. Imagery of a victimized self, in association with self-punishment and injustice collection (Bergler), leads to efforts to induce guilty and reparative behaviors toward oneself. The helpless and infantile self gives rise to behavioral expressions that attract care and thereby neutralize the object's hostility and that of the wider social surround. Thus, the inner self pictures itself in one or another safe position, which would incentivize the self's efforts to engage an external derivative of the primary object, to attain this object's recognition and assurances (**see Figure below**).

Exhibitionistic behaviors (and hence also ambition) may be related to separation calls, given that the purpose of exhibitionism is to attract attention (specifically from the primary object). Once separation anxiety has been superseded, developmentally and evolutionarily, by the anxiety that is associated with the realization of one's separateness, a separateness that is aversive because it bears within it a sense of vulnerability to being attacked and annihilated by others, the task set before the individual is to attract *positive* attention (narcissistically nourishing attention). Attention from and approach by the object calms separation anxiety; and, if anxiety is intense, negative attention from the object would be preferable to no attention, in which case there could be said to be a regression to a state when attention received by the infant was not differentiated into positive and negative attention. Such regression, manifesting as primitive attention-seeking behaviors, may occur when anxiety is intense (and self-disintegration [Kohut] is imminent). When anxiety is less intense, the capacity of foresight (anticipation) can be engaged, wherein the self is viewed in a desired safe position ('idealized image'), so as to provide guidance for adaptive goal-directed behavior. When the group is not clearly centered on a leader and the group's aggressive potential is not clearly bound to an external objective, the need increases to appease others within the group and to ensure oneself of the benevolence of whatever transiently occupies the role of the primary object. It is then that the self (the inner self) emerges as a defensive entity that guides efforts to enhance the individual's acceptability and approvability in the eyes of the superego internally or externally. Being integrated into a cohesive group and identified with a common cause, on the other hand, is associated with regression in superego development (Freud) and dissolution of the self (loss of one's awareness of oneself as an individual). In a cohesive group, the narcissistic balance of each individual would be upheld by the occupation (and defense) of a relatively stable position in the social hierarchy (centered on a leader), through the exchange of signals of submission and dominance with others in the group. There will be an equilibrium in the group between expressions of dominance by some and expressions of submission or subservience by others, affording each individual with a degree of security (protection against others' innate hostility) and maximizing the amount of safety distributed across the group (and hence stabilizing the group). Dominance positions, being dynamically maintained in such a network, are of vital importance to each member because they define each member's closeness to the leader, the ultimate source of narcissistic gratification and provider of safety.

Aggression, compliance, and the ability to channel narcissistic demands (demands for attention) into realistic directions are employed in shaping the selfobject surround (representing the availability of selfobject responses, i.e. of narcissistic supplies). Aggression and compliance, in particular, are used to control narcissistic resources (the selfobject surround), which are developmentally continuous with the availability and responsiveness of the mother, *in much the same way as territorial animals manage their territory. The exhibitionistic component of behavior is more directly concerned with attaining positive attention (narcissistic supplies) from an external representative of the primary object (which is, at the same time, a projection of the superego).*

Narcissistic homeostasis (self-cohesion) is upheld proximally by soliciting approval and admiration from the representative of the primary object and distally (or more abstractly) by defending one's social position (using aggression and submission) or by enhancing one's approvability within the group and in the eyes of the leader (so as to control access to narcissistic resources and their responsiveness to narcissistic demands). The self of the 'reality'-oriented social actor and observer serves as a point of reference to narcissistic resources and encapsulates rights of access to them, whereby self-esteem is the confident expectation of others' self-confirming responses to one's exhibitionistic and care-seeking displays. The self can also be said to be a distillate of received narcissistic feedback and of environmental cues relating to the availability of such feedback (cues which in turn are controlled by the social actor, in part through acquisition of prestige and possessions). While the self (or 'ego', for the purpose of this book) is situated on the margins of conscious experience of the external world, the ego ideal (ideal self) can be found in the margins of internal imagery. The ego ideal (ideal self) can adopt various forms, any of which can be used, in states of detachment, to solicit narcissistic supplies from an imaginary internal audience (the superego) or, in 'reality'-oriented states, to set goals for actions, actions that in themselves express various combinations of assertive, compliant, exhibitionistic, and care-seeking impulses. Thus, while the ego ideal is situated vis-à-vis the superego in the realm of imagery, the self is situated vis-à-vis the superego project in the 'real' world; and while the ego ideal serves anticipatory functions, the self serves functions related to self-localization in the social landscape (emphasizing the suggested derivation of goal-directed social behavior from evolutionarily older goal-directed locomotor behavior [Behrendt, 2015]).

References

Adler, A. (1927/1998). *Understanding Human Nature.* Oxford: Oneworld Publications.

Adler, A. (1938/1998). *Social Interest.* Oxford: Oneworld Publications.

Adler, A. (1965). *Superiority and Social Interest: A collection of later writings* (H.L. Ansbacher & R.R. Ansbacher, eds.). London: Routledge and Kegan Paul.

Arieti, S. (1970). The structural and psychodynamic role of cognition in the human psyche. In: S. Arieti (ed.), *The World Biennial of Psychiatry and Psychotherapy* (Vol. 1, pp. 3-34). New York: Basic Books.

Arieti, S. (1973). The interpersonal and the intrapsychic in severe psychopathology. In: E.G. Witenberg (ed.), *Interpersonal Explorations in Psychoanalysis: New Directions in Theory and Practice* (pp. 120-131). New York: Basic Books.

Arlow, J.A. (1989). Psychoanalysis and the quest for morality. In: H.P. Blum, E.M. Weinshel, & F.R. Rodman (eds.), *The Psychoanalytic Core: Essays in Honor of Leo Rangell, M.D.* (pp. 147-166). Madison, Connecticut: International Universities Press.

Behrendt, R.P. (2015). *Narcissism and the Self: Dynamics of Self-Preservation in Social Interaction, Personality Structure, Subjective Experience, and Psychopathology.* London: Palgrave Macmillan.

Bergler, E. (1949). *The Basic Neurosis: Oral Regression and Psychic Masochism.* New York: Grune & Stratton.

Bergler, E. (1952). *The Superego: Unconscious Conscience.* New York: Grune & Stratton.

Berkowitz, L. (1989). Laboratory experiments in the study of aggression. In: J. Archer & K. Browne (eds.), *Human Aggression: Naturalistic Approaches.* London: Routledge.

Bion, W.R. (1952/1980). Group dynamics: A re-view. In: S. Scheidlinger (ed.), *Psychoanalytic Group Dynamics: Basic Readings* (pp. 77-107). New York: International Universities Press.

Bion, W.R. (1962). The psycho-analytic study of thinking: II. A theory of thinking. *International Journal of Psychoanalysis* 43: 306-310.

Blanchard, D.C. & Blanchard, R.J. (1989). Experimental animal models of aggression: what do they say about human behaviour. In: J. Archer & K. Browne (eds.), *Human Aggression: Naturalistic Approaches.* London: Routledge.

Bowlby, J. (1973). *Separation: Anxiety and Anger.* New York: Basic Books.

Brandchaft, B. (1985). Resistance and defense: An intersubjective view. In: A. Goldberg (ed.), *Progress in Self Psychology* (Vol. 1, pp. 88-96). New York: Guilford Press.

Bursten, B. (1973). Some narcissistic personality types. *International Journal of Psychoanalysis* 54: 287-300.

Cashdan, S. (1988). *Object Relations Therapy: Using the Relationship.* New York: W.W. Norton & Company.

Eibl-Eibesfeldt, I. (1970/1971). *Love and Hate: On the Natural History of Basic Behaviour Patterns.* London: Methuen & Co.

Erikson, E.H. (1950/1977). *Childhood and Society.* London: Paladin Grafton Books.

Erikson, E.H. (1959/1980). Ego development and historical change. In: S. Scheidlinger (ed.), *Psychoanalytic Group Dynamics: Basic Readings* (pp. 189-212). New York: International Universities Press.

Fairbairn, W.R.D. (1952). *Psychoanalytic Studies of the Personality.* London: Routledge & Kegan Paul.

Federn, P. (1952). *Ego Psychology and the Psychoses* (ed., E. Weiss). New York: Basic Books.

Fenichel, O. (1946). *The Psychoanalytic Theory of Neurosis.* London: Routledge & Kegan Paul.

Flugel, J.C. (1945). *Man, Morals and Society.* London: Duckworth.

Freud, A. (1937/1966). *The ego and the mechanisms of defense.* New York: International Universities Press.

Freud, S. (1914/1957). On narcissism: An introduction. In: J. Strachey (ed.), *The Standard Edition of the Complete Psychological Works of Sigmund Freud* (Vol. 14, pp. 67-102). London: The Hogarth Press.

Freud, S. (1917/1957). Mourning and melancholia. In: J. Strachey (ed.), *The Standard Edition of the Complete Psychological Works of Sigmund Freud* (Vol. 14, pp. 237-258). London: The Hogarth Press.

Freud, S. (1921/1922). *Group Psychology and the Analysis of the Ego* (tr., J. Strachey). London: The Hogarth Press.

Freud, S. (1923/1961). The ego and the id. In: J. Strachey (ed.), *The Standard Edition of the Complete Psychological Works of Sigmund Freud* (Vol. 19, pp. 12-66). London: The Hogarth Press.

Freud, S. (1930/2002). *Civilisation and Its Discontents* (tr., D. McLintock). London: Penguin.

Goodman, G. (2002). *The Internal World and Attachment.* Hillsdale, New Jersey: The Analytic Press.

Greenson, R.R. (1947/1978). On gambling. In: R.R. Greenson, *Explorations in Psychoanalysis* (pp. 1-15). New York: International Universities Press.

Greenson, R.R. (1958/1978). On screen defenses, screen hunger, and screen identity. In: R.R. Greenson, *Explorations in Psychoanalysis* (pp. 111-132). New York: International Universities Press.

Greenson, R.R. (1962/1978). On enthusiasm. In: R.R. Greenson, *Explorations in Psychoanalysis* (pp. 171-189). New York: International Universities Press.

Greenson, R.R. (1973/1978). The personal meaning of perfection. In: R.R. Greenson, *Explorations in Psychoanalysis* (pp. 479-490). New York: International Universities Press.

Hartmann, H. (1964). *Essays on Ego Psychology: Selected Problems in Psychoanalytic Theory.* London: The Hogarth Press.

Hass, H. (1968/1970). *The Human Animal: The Mystery of Man's Behaviour.* London: Hodder and Stoughton.

Heard, D.H. & Lake, B. (1986). The attachment dynamic in adult life. *British Journal of Psychiatry* 149: 430-438.

Hendrick, I. (1958). *Facts and Theories of Psychoanalysis*, 3rd edition. New York: Alfred A. Knopf.

Horney, K. (1937). *The Neurotic Personality of our Time.* New York: W.W. Norton & Company.

Horney, K. (1939). *New Ways in Psychoanalysis.* New York: W.W. Norton & Company.

Horney, K. (1945/1992). *Our Inner Conflicts.* New York: W.W. Norton & Company.

Horney, K. (1950/1991). *Neurosis and Human Growth: The Struggle Toward Self-Realization.* New York: W.W. Norton & Company.

Jacobson, E. (1964). *The Self and the Object World.* New York: International Universities Press.

Joffe, W.G. & Sandler, J. (1965/1987). Pain, depression, and individuation. In: J. Sandler (ed.), *From Safety to Superego: Selected Papers of Joseph Sandler* (pp. 154-179). New York: Guilford Press.

Joffe, W.G. & Sandler, J. (1967/1987). On disorders of narcissism. In: J. Sandler (ed.), *From Safety to Superego: Selected Papers of Joseph Sandler* (pp. 180-190). New York: Guilford Press.

Joffe, W.G. & Sandler, J. (1968/1987). Adaptation, affects, and the representational world. In: J. Sandler (ed.), *From Safety to Superego: Selected Papers of Joseph Sandler* (pp. 221-234). New York: Guilford Press.

Joseph, B. (1986). Envy in everyday life. *Psychoanalytic Psychotherapy* 2: 13-22.

Kernberg, O.F. (1970). Factors in the psychoanalytic treatment of narcissistic personalities. *Journal of the American Psychoanalytic Association* 18: 51-85.

Kernberg, O.F. (1992). *Aggression in Personality Disorders and Perversions*. New Haven, Connecticut: Yale University Press.

Kernberg, O.F. (1996). A psychoanalytic theory of personality disorders. In: J.F. Clarkin & M.F. Lenzenweger (eds.), *Major Theories of Personality Disorder* (pp. 106-140). New York: Guilford Press.

Klein, M. (1932/1937). *The Psycho-Analysis of Children*. London: The Hogarth Press & The Institute of Psycho-Analysis.

Klein, M. (1937). Love, Guilt and Reparation. In: M. Klein & J. Riviere, *Love, Hate and Reparation* (pp. 57-119) (Psycho-Analytical Epitomes No. 2). London: The Hogarth Press and the Institute of Psycho-Analysis.

Klein, M. (1940/1948). Mourning and its relation to manic depressive states. In: M. Klein, *Contributitons to Psycho-Analysis, 1921-1945*. London: Hogarth.

Klein, M. (1946/1952). Notes on some schizoid mechanisms. In: M. Klein, P. Heimann, S. Isaacs, & J. Riviere (eds.), *Developments in Psycho-Analysis* (pp. 292-320). London: The Hogarth Press.

Kohut, H. (1966). Forms and transformations of narcissism. *Journal of the American Psychoanalytic Association* 14: 243-272.

Kohut, H. (1971). *The Analysis of the Self: A Systematic Approach to the Psychoanalytic Treatment of Narcissistic Personality Disorders*. New York: International Universities Press.

Kohut, H. (1977). *The Restoration of the Self*. New York: International Universities Press.

Kohut, H. (1983). Selected problems of self psychological theory. In: J.D. Lichtenberg & S. Kaplan (eds.), *Reflections on Self Psychology* (pp. 387-416). Hillsdale, New Jersey: The Analytic Press.

Kohut, H. (1984). *How Does Analysis Cure?*, ed. A. Goldberg with P. Stepansky. Chicago & London: The University of Chicago Press.

Kohut, H. & Wolf, E.S. (1978). The disorders of the self and their treatment: An outline. *International Journal of Psychoanalysis* 59: 413-425.

Laing, R.D. (1960). *The Divided Self*. London: Tavistock Publications.

Laughlin, H.P. (1970). *The Ego and Its Defenses*. New York: Appleton-Century-Crofts.

Lebovici, S. (1989). Precocious aspects of the formation of morality. In: H.P. Blum, E.M. Weinshel, & F.R. Rodman (eds.), *The Psychoanalytic Core: Essays in Honor of Leo Rangell, M.D.* (pp. 421-434). Madison, Connecticut: International Universities Press.

Lorenz, K. (1963/2002). *On Aggression*. London: Routledge.

Lorenz, K. (1973/1977). *Behind the Mirror: A Search for a Natural History of Human Knowledge*. London: Methuen & Co.

Mahler, M.S. (1967). On human symbiosis and the vicissitudes of individuation. *Journal of the American Psychoanalytic Association* 15: 740-763.

Mahler, M.S. (1972). On the first three subphases of the separation-individuation process. *International Journal of Psychoanalysis* 53: 333-338.

McCluskey, U. (2002). The dynamics of attachment and systems-centered group psychotherapy. *Group Dynamics: Theory, Research, and Practice* 6: 131-142.

Miller, A. (1979). Depression and grandiosity as related forms of narcissistic disturbances. *International Journal of Psychoanalysis* 60: 61-67.

Modell, A.H. (1975). A narcissistic defence against affects and the illusion of self-sufficiency. *International Journal of Psychoanalysis* 56: 275-282.

Money-Kyrle, R.E. (1961/1978). *Man's Picture of his World: A Psycho-analytic Study*. London: Duckworth.

Morrison, A.P. (1983). Shame, ideal self, and narcissism. *Contemporary Psychoanalysis* 19: 295-318.

Moses, R. (1989). Shame and entitlement: Their relation to political process. In: H.P. Blum, E.M. Weinshel, & F.R. Rodman (eds.), *The Psychoanalytic Core: Essays in Honor of Leo Rangell, M.D.* (pp. 421-434). Madison, Connecticut: International Universities Press.

Moynihan, M.H. (1998). *The Social Regulation of Competition and Aggression in Animals*. Washington: Smithsonian Institution Press.

Nunberg, H. (1955). *Principles of Psychoanalysis*. New York: International Universities Press.

Rado, S. (1928/1956). The problem of melancholia. In: *Psychoanalysis of Behavior: Collected Papers* (pp. 47-63). New York: Grune & Stratton.

Rado, S. (1956). *Psychoanalysis of Behavior: Collected Papers*. New York: Grune & Stratton.

Reddy, V. (2003). On being the object of attention: Implications for self-other consciousness. *Trends in Cognitive Sciences* 7: 397-402.

Redl, F. (1942/1980). Group emotion and leadership. In: S. Scheidlinger (ed.), *Psychoanalytic Group Dynamics: Basic Readings* (pp. 15-68). New York: International Universities Press.

Reich, A. (1953). Narcissistic object choice in women. *Journal of the American Psychoanalytic Association* 1: 22-44.

Reich, A. (1960). Pathologic forms of self-esteem regulation. *The Psychoanalytic Study of the Child* 15: 215-232.

Reich, W. (1928/1950). On character analysis. In: R. Fliess (ed.), *The Psychoanalytic Reader: An Anthology of Essential Papers with Critical Introductions* (pp. 106-123). London: The Hogarth Press.

Reich, W. (1929/1950). The genital character and the neurotic character. In: R. Fliess (ed.), *The Psychoanalytic Reader: An Anthology of Essential Papers with Critical Introductions* (pp. 124-144). London: The Hogarth Press.

Reiser, D.E. (1986). Self psychology and the problem of suicide. In: A. Goldberg (ed.), *Progress in Self Psychology* (Vol. 2, pp. 227-241). New York: Guilford Press.

Rickles, W.H. (1986). Self psychology and somatization: An integration with alexithymia. In: A. Goldberg (ed.), *Progress in Self Psychology* (Vol. 2, pp. 212-226). New York: Guilford Press.

Riviere, J. (1936). A contribution to the analysis of the negative therapeutic reaction. *International Journal of Psychoanalysis* 17: 304-320.

Riviere, J. (1937). Hate, greed and aggression. In: M. Klein & J. Riviere, *Love, Hate and Reparation* (pp. 3-53) (Psychoanalytic Epitomes No. 2). London: The Hogarth Press and the Institute of Psycho-Analysis.

Rosenfeld, H.A. (1965). *Psychotic States: A Psychoanalytical Approach.* London: Maresfield Reprints.

Rothstein, A. (1979). The theory of narcissism: An object-relations perspective. *The Psychoanalytic Review* 66: 35-47.

Sandler, J. (1960a/1987). The background of safety. In: J. Sandler (ed.), *From Safety to Superego: Selected Papers of Joseph Sandler* (pp. 1-8). New York: Guilford Press.
[originally published in: *International Journal of Psychoanalysis* 41: 352-356]

Sandler, J. (1960b/1987). The concept of superego. In: J. Sandler (ed.), *From Safety to Superego: Selected Papers of Joseph Sandler* (pp. 17-44). New York: Guilford Press.

Sandler, J. (1985). *The analysis of defense,* With A. Freud. New York: International Universities Press.

Sandler, J. (1989). Unconscious wishes and human relationships. In: J. Sandler (ed.), *Dimensions of Psychoanalysis* (pp. 65-81). Madison, Connecticut: International Universities Press.

Sandler, J. & Joffe, W.G. (1968/1987). Psychoanalytic psychology and learning theory. In: J. Sandler (ed.), *From Safety to Superego: Selected Papers of Joseph Sandler* (pp. 255-263). New York: Guilford Press.

Sandler, J. & Nagera, H. (1963/1987). The metapsychology of fantasy. In: J. Sandler (ed.), *From Safety to Superego: Selected Papers of Joseph Sandler* (pp. 90-120). New York: Guilford Press.

Sandler, J. & Sandler, A.-M. (1978). On the development of object relations and affects. *International Journal of Psychoanalysis* 59: 285-296.

Sandler, J., Holder, A., & Meers, D. (1963/1987). Ego ideal and ideal self. In: J. Sandler (ed.), *From Safety to Superego: Selected Papers of Joseph Sandler* (pp. 73-89). New York: Guilford Press.

Schachtel, E.G. (1973). On attention, selective inattention, and experience: an inquiry into attention as an attitude. In: E.G. Witenberg (ed.), *Interpersonal Explorations in Psychoanalysis: New Directions in Theory and Practice* (pp. 40-66). New York: Basic Books.

Schafer, R. (1997). Conformity and individualism. In: E.R. Shapiro (ed.), *The Inner World in the Outer World* (pp. 27-42). New Haven, Connecticut: Yale University Press.

Schecter, D.E. (1973). On the emergence of human relatedness. In: E.G. Witenberg (ed.), *Interpersonal Explorations in Psychoanalysis: New Directions in Theory and Practice*, (pp. 17-39). New York: Basic Books.

Schecter, D.E. (1978). Attachment, detachment, and psychoanalytic therapy. In: E.G. Witenberg (ed.), *Interpersonal Psychoanalysis: New Directions* (pp. 81-104). New York: Gardner Press.

Scheidlinger, S. (1964/1980). Identification, the sense of belonging and of identity in small groups. In: S. Scheidlinger (ed.), *Psychoanalytic Group Dynamics: Basic Readings* (pp. 213-231). New York: International Universities Press.

Scheidlinger, S. (1968/1980). The concept of regression in group psychotherapy. In: S. Scheidlinger (ed.), *Psychoanalytic Group Dynamics: Basic Readings* (pp. 233-254). New York: International Universities Press.

Schilder, P. (1951). *Psychoanalysis, Man, and Society.* New York: W.W. Norton & Company.

Schilder, P. (1976). *On Psychoses* (ed., L. Bender). New York: International Universities Press.

Schultz-Henke, H. (1951/1988). *Lehrbuch der Analytischen Psychotherapie.* Stuttgart: Georg Thieme Verlag.

Segal, H. (1973). *Introduction to the work of Melanie Klein.* London: Hogarth Press.

Shapiro, D. (2000). *Dynamics of character.* New York: Basic Books.

Silverman, D.K. (1991). Attachment patterns and Freudian theory: An integrative proposal. *Psychoanalytic Psychology* 8: 169-193.

Spillius, E.B. (1993). Varieties of envious experience. *International Journal of Psychoanalysis* 74: 1199-1212.

Stolorow, R.D. (1983). Self psychology – a structural psychology. In: J.D. Lichtenberg & S. Kaplan (eds.), *Reflections on Self Psychology* (pp. 287-296). Hillsdale, New Jersey: The Analytic Press.

Stolorow, R.D. (1985). Toward a pure psychology of inner conflict. In: A. Goldberg (ed.), *Progress in Self Psychology* (Vol. 1, pp. 193-201). New York: Guilford Press.

Storr, A. (1968). *Human Aggression*. New York: Atheneum.

Tolpin, P.H. (1986). What makes for effective analysis? In: A. Goldberg (ed.), *Progress in Self Psychology* (Vol. 2, pp. 95-105). New York: Guilford Press.

Wallerstein, R.S. (1983). Self psychology and classical psychoanalytic psychotherapy: The nature of their relationship. In: A. Goldberg (ed.), *The Future of Psychoanalysis* (pp. 19-63). New York: International Universities Press.

Wilson, S.L. (1985). The self-pity response: A reconsideration. In: A. Goldberg (ed.), *Progress in Self Psychology* (Vol. 1, pp. 178-190). New York: Guilford Press.

Winnicott, D.W. (1958/1965). The capacity to be alone. In: D.W. Winnicott, *The Maturational Processes and the Facilitating Environment* (pp. 29-36). New York: International Universities Press.

Winnicott, D.W. (1960b/1965). Ego distortion in terms of true and false self. In: D.W. Winnicott, *The Maturational Processes and the Facilitating Environment* (pp. 140-152). New York: International Universities Press.

Winnicott, D.W. (1989). *Psycho-Analytic Explorations* (C. Winnicott, R. Shepherd, & M. Davis., eds.). Cambridge, MA: Harvard University Press.

Wolf, E. (1988). *Treating the Self: Elements of Clinical Self Psychology*. New York: Guilford Press.

Index

W

www.ingramcontent.com/pod-product-compliance
Lightning Source LLC
Chambersburg PA
CBHW050515280326
41932CB00014B/2326